ON UNDERSTANDING PHYSICS

ON UNDERSTANDING PHYSICS

by

W. H. WATSON

M.A., Ph.D. (Edin.), Ph.D. (Cantab.), F.R.S.C.

Assistant Professor of Physics, McGill University, formerly Lecturer in Natural Philosophy, Edinburgh University

CAMBRIDGE
AT THE UNIVERSITY PRESS
1938

CAMBRIDGE
UNIVERSITY PRESS

University Printing House, Cambridge CB2 8BS, United Kingdom

Cambridge University Press is part of the University of Cambridge.

It furthers the University's mission by disseminating knowledge in the pursuit of education, learning and research at the highest international levels of excellence.

www.cambridge.org
Information on this title: www.cambridge.org/9781316601839

First published 1938
First paperback edition 2015

A catalogue record for this publication is available from the British Library

ISBN 978-1-316-60183-9 Paperback

And therefore most times, is the poverty of human understanding copious in words, because enquiring hath more to say than discovering....

Augustine, 'Confessions'

CONTENTS

PREFACE

THIS BOOK is based on a course of lectures on the logic of physics given during the last few years to graduate students of physics at McGill University and is offered in the hope that the interest of physicists in particular and scientists in general may be drawn to developments in modern philosophy which promise to be of great importance to learning. These developments are largely due to Dr Ludwig Wittgenstein. Though his writings and teaching have commanded the attention of those interested in mathematical philosophy (to use Russell's term), they are certainly little known by experimental scientists. Wittgenstein's *Tractatus Logico-Philosophicus* is hardly the book one would expect to attract the interest of working scientists, for in order to understand it one really requires the aid of philosophical discussion to explain the thoughts that are expressed in it, and besides, it has not been understood even by some professional logicians. In his Cambridge lectures since 1929, Wittgenstein's method of exposition has necessarily differed from that of the *Tractatus*, quite apart from any changes of view from what is expressed there; much of his teaching is by means of discussions. This seems to be the right way to teach philosophy; a book is of use only as a source of topics for discussion.

During the years 1929–31, I attended Wittgenstein's lectures in Cambridge and recently have had the pleasure of reading in manuscript, sent to me by Dr Wittgenstein, his lectures given during the session 1933–4. The present book is not intended in any way as a report of these lectures, but naturally it bears evidence of their influence, and also of the influence of private conversation with the man whose friendship I am happy to possess. Hoping that the general philosophical reader will in the near future have the opportunity to consult Wittgenstein's own writings, I have hardly dealt at all with questions that naturally belong to the philosophy class room which is the proper place to wage 'the fight against the fascination which forms of expression have over us'. My main interest is physics, and what is written here presents the results of my reflections in clearing up my own understanding of physics. Some of my readers are likely to be irritated by the form of presenting the subject-matter of this book, and will probably feel that a book on the philosophy of physics should resemble a book on mechanics, beginning with 'the simplest conceptions' and working up to 'more complicated ideas'. If in the course of reading this book they are disabused of this prejudice, they will be freed from a very cramping form of thinking philosophically about science. The form chosen should make it clear that there is no question of a textbook or a treatise, and equally that this book is intended primarily for readers who are familiar with physics.

It is widely recognised by leading physicists that the training of young physicists in our universities should not be limited to modern knowledge but that senior students should be brought to interest themselves in the history of their subject—of the men and of the development of ideas. On the other hand, although it must be admitted that those in favour are enthusiastic, there is no such unanimity about the need for developing interest in the philosophy of physics nor indeed agreement as to the meaning of the term. It is unfortunate, therefore, that history and philosophy should be grouped together in connection with science courses at some universities. The only justification for the association of philosophy with science in the training of scientists is that it is of some definite use—that there is some technical value in the intellectual discipline of philosophising. Indeed, one can advance the weightiest of reasons for discouraging the training of a scientist in traditional philosophy, namely, that it will unfit him for the work of scientific investigation either by encouraging wild speculation, and developing literary facility in its expression, or by making him an expert in 'logic chopping'. It is because scientists can be provided with a method in approaching the criticism of their subject, that philosophy in the sense of Wittgenstein (or as Wittgenstein puts it, "that activity which is one of the heirs of what was once called philosophy") has a proper technical value for scientists. From the point of view of training scientists, the most important task is to show the limits of achievement to be expected of philosophy. Not only what is beyond the scope of philosophy must be indicated, but that some definite goal can be attained by philosophical activities. In this way we should gradually eliminate from the outlook of scientists two points of view that do little credit to the profession: first, the view that philosophy can accomplish nothing of value, and second, toleration of the type of professedly philosophical writing which accompanies popular expositions of modern science, under the guise of synthesising knowledge, for some religious end.

I am well aware that in maintaining that it is necessary in the first place to show the limits of achievement to be expected in philosophy, one is certain to incur reproof at the hands of not a few professional philosophers. It may seem an impertinence on my part to write boldly on matters which some of my philosophical friends wish to preserve intact in their traditional muddle. Perhaps it is forgotten, however, that the great philosophers of the past always had in mind the solution of the problems they raised, and would be appalled at the thought that they had produced an indestructible cud for academic rumination. Surely we can settle philosophical problems now, if questions that cannot be settled now do not belong to philosophy; and when we ask what we mean by the words we use, we have the right to expect a clear explanation —although not necessarily a simple one. This is the sense in which I approached these lectures on philosophy for physicists, and I offer no apology for having ignored nearly completely the history of philosophy and the modern writings of scientists on the philosophy of science. The great value to physicists in coming to understand Wittgenstein's method, even if they are unversed in the technique of symbolic logic, is that they will become clear about symbolism in general and consequently more clear about physics itself. Probably no subject is more badly taught than mechanics in its introductory stages, that is, whenever it is a question of explaining the method of representation to be employed. To remove the confusion of thought to which this is due is one of the important objects of training some physicists in philosophy. It is a necessary prelude to any attempt to eliminate obvious intellectual discomfort that has been associated with the recent development of the subject to describe atomic processes.

Thus, on the one hand, I have aimed at interesting physicists in philosophy in Wittgenstein's sense, on the other I have attempted to show its value for clarifying physics—especially mechanics which is the logical backbone of the subject. In this book are discussed among others some of the logical problems that have been thrown into prominence during the development of atomic mechanics, and the reader will find that the treatment here differs from that to be found in the writings of mathematical physicists and others who have dealt with them. The important step is that these matters should be regarded as logical problems; one might say that whether the analysis is complete or not will show itself in the same way as correctness or error in a mathematical theory, but of course the use of language is a more complicated affair than mathematics.

W. H. WATSON

MONTREAL
March 1938

CHAPTER I

DISCIPLINE IN PHILOSOPHY

No physicist who has made personal contact with the Cavendish Laboratory will have difficulty in recognising the phrase 'Get on with it' which has served to stimulate many research workers there. Sometimes the words have acted as a goad to irritate still further a man who has seen many unfruitful hours pass in his struggle with recalcitrant matter. But, apart from its private implications, this simple English sentence sums up a point of view which does not readily translate itself into the entertaining literature of popular science and speculative philosophy. 'Getting on with it' implies not only a definite job to be accomplished and a method for doing it, it insists that all activities which do not contribute to the speedy accomplishment of the task are to be firmly set aside. It emphasises the professional point of view as opposed to the dilettante's. It does not aim at discouraging thought and reflection, but it does have in mind the suppression of speculation which is not directed towards immediate experimental investigation. The work of the physicist is to find what happens in the world, and he is not doing that work when he engages in metaphysical speculation or indulges in omnibus discussions, however pleasant these may be. Not unreasonably a builder does not wish to see his bricklayers engaged in talking politics when they ought to be laying bricks, and with the same point, a director of research is interested to have the physicists under his charge lay the bricks of the structure of their science. The physicist, however, is not in the happy position of the bricklayer in that no one takes the trouble to assert the importance of bricklaying for philosophy or of philosophy for bricklaying, whereas at the present time the interest of philosophers in physics and of physicists in philosophy is greater than at any time in the last two hundred years. Indeed, at the end of last century, with very few exceptions, the attitude of physicists was one of indifference, if not contempt, for metaphysical inquiry. This temper of mind was strongly condemned by Hertz who wrote, in the Introduction to his *Principles of Mechanics* (p. 23):

> A doubt which makes an impression on our mind cannot be removed by calling it metaphysical; every thoughtful mind has needs which scientific men are accustomed to denote as metaphysical.

To-day, physicists and mathematicians have produced much philosophical literature which, however, is in striking contrast to the attitude we have just noticed of the professional workers in their laboratories.

Of course, when one examines the philosophical prejudices of physicists and mathematicians one finds a great variety of belief and soon discovers that not all of them attach much importance to their own particular point of view. For every young man who is seriously interested in philosophical questions, there are perhaps ten, representative of all ages, who have an attitude of benevolent indifference and toleration. Most of the older scientists who are interested in philosophy, however, zealously uphold one of the views, either, that physics helps the philosopher to present a picture of the world which is important in the religious sense, or that philosophy is somehow the basis of mathematics and physics. The existence of this diversity of opinion makes one doubt whether the subject-matter which engages the attention of professional philosophers has anything at all to do with physics and mathematics. How can it be said that philosophy, as traditionally understood, is of technical value to scientists in their everyday work when it seems to be the case that the majority of them rarely give philosophy a moment's thought? If philosophy were of technical value to physicists, it is reasonable to suppose that physicists would not be able to function until philosophy, in so far as it affects physics, had been summarily dealt with, for philosophical problems that are technically important for physicists must be solved in the course of the development of physics. In the history of physics there have actually been instances of this process, but in each case the philosophical atmosphere which at first pervaded the controversy has dispersed in time, and the final achievement is now seen as a modification or clarification of physical theory. The history of the changes which have brought about the clear statement of theory may be forgotten without loss to the student subsequently learning the subject for the first time. In this respect the natural history of physical theories is mirrored in the history of all branches of knowledge, and in the experience of individuals. If I am puzzled in learning the differential calculus, because I have not yet clearly formed the ideas to be learned, it is not necessary for me to recall that confusion at a later time when I have become clear about the matter and wish to use my knowledge. When one is puzzled one is aware of problems which one does not know how to solve although it seems essential to do so. These problems set by the sense of unclearness are not solved but removed when the source of confusion of thought has been discovered, and one really understands.

When new theories arise in physics they do not always appear with the clarity which is later imposed on them by refinement of the ideas and removal of misleading or contradictory expressions in the exposition of them. The questions which show the need for clarifying the theory are essentially philosophical questions, although they will only be recognised as such and engage the interest of philosophers if they are simple enough and of sufficiently general importance. For example, one would not usually call philosophical the problem that arises in the statement of Faraday's law of electromagnetic induction,* but one does refer to the philosophical problem of relative motion.

Philosophical problems, generally speaking, concern

* Faraday's law is often stated in terms of the rate of change of the number of lines of magnetic induction linked with the circuit in which the e.m.f. is induced. It is puzzling to recognise just where this statement differs in effect from that in terms of number of lines cut per second by the circuit, and what exactly is meant by linkage.

matters with which everyone is acquainted. For this reason, perhaps, there is an appearance of triviality about them. Nevertheless, it is also true to say that one is impressed by a philosophical problem; I do not refer to impressiveness resulting from literary skill in stating it, by which effects are produced that belong to the art of the orator and are out of place in arduous inquiry. In order to show what a philosophical problem is and how simply it may be stated, one cannot do better than turn to the words of Augustine who expressed with magnificent directness how he was puzzled about 'time'. The following is Pusey's translation, which hardly does justice to the essential modernness of the original.

For what is time? Who can readily and briefly explain this? Who can even in thought comprehend it, so as to utter a word about it? But what in discourse do we mention more familiarly and knowingly than time? And, we understand, when we speak of it; we understand also when we hear it spoken of by another. What then is time? If no one asks me I know; if I wish to explain it to one that asketh, I know not; yet I say boldly, that I know, that if nothing passed away, time past were not; and if nothing were coming, a time to come were not; and if nothing were, time present were not. Those two times then, past and to come, how are they, seeing the past now is not, and that to come is not yet? But the present, should it always be present and never pass into time past, verily it should not be time, but eternity.

Confessions of St Augustine, XIV, 17.

Philosophical problems differ from ordinary physical problems in this respect: we have a method by which to solve the physical problem, and when the problem is stated we know what kind of answer to expect; but the philosophical problem solves itself when we are shown that we were mistaken in the answer we expected, and are forced to clear up the grammar of our language. Hertz clearly analysed such a situation in connection with the philosophical question as to the nature of electricity.

Weighty evidence seems to be furnished by the statement which one hears with wearisome frequency, that the nature of force is a mystery, that one of the chief problems of physics is the investigation of the nature of force and so on. In the same way electricians are continually attacked as to the nature of electricity. Now, why is it that people never in this way ask what is the nature of gold, or what is the nature of velocity? Is the nature of gold better known to us than that of electricity, or the nature of velocity better than that of force? Can we by our conceptions, by our words, completely represent the nature of anything? Certainly not. I fancy the difference must lie in this. With the terms 'velocity' and 'gold' we connect a large number of relations to other terms, and between all these relations we find no contradictions which offend us. We are therefore satisfied and ask no further questions. But we have accumulated around the terms 'force' and 'electricity' more relations than can be completely reconciled amongst themselves. We have an obscure feeling of this and want to have things cleared up. Our confused wish finds expression in the confused question as to the nature of force and electricity. *But the question is mistaken with regard to the answer which it expects.* It is not by finding out more and fresh relations and connections that it can be answered but by removing the contradictions existing between those already known, and thus perhaps by reducing their number. When these painful contradictions are removed, the question as to the nature of force will not have been answered; but our minds no longer vexed will cease to ask illegitimate questions.*

In making this comparison between an ordinary scientific problem and a philosophical problem we have given a special meaning to the term 'philosophy', emphasising the critical as opposed to the speculative activity of philosophers. Another way of stating the point of view is that the problems of philosophy are logical problems only, and that metaphysical inquiry is based on misunderstanding of the possibilities of language. To assert this is not to undervalue the work of the great metaphysicians of the past, but to point out the road along which progress is being made in modern philosophy. One might say that philosophy is emerging from the alchemy stage of asking for something it *cannot* have, into the disciplined age which corresponds to the era of chemical science, when the chemist, instead of trying to transmute base metals

* Introduction to *The Principles of Mechanics*, pp. 7–8.

into gold, is content to study them all and find their proper uses. The importance which has always been attached to philosophical questions is being attached now in a new way. It is no longer a matter of the halo of mysticism and of its intellectual counterpart, the unification of all knowledge, but is just the kind of importance which a carpenter sees in having his tools sharp or which an engineer has in mind when he says it is important to know how this or that machine works. It is the importance which every physicist feels in understanding clearly the set of ideas he has to use in describing physical phenomena.

In philosophy there are no physical facts which the teacher has to convey to the students under his guidance: whereas in physics, the student must be told what others have found to happen in the world. The student of philosophy already knows how to speak in the manner that is understood by his fellows in everyday affairs. When he begins philosophy, questions are asked which he can answer without learning new facts about the world. It is true the teacher may have difficulty in getting him to see clearly distinctions to which he had not previously given attention. Just as there are many things which we pass by the roadside without attending to them, but which we cannot fail to observe when we are told to look, so there are logical distinctions, prohibitions and possibilities which are exhibited to us when we reflect on the use of language, and it may often help to have someone give the hint where to look.

It must be the nature of philosophical problems that their solution does not have to await the becoming of facts. It must be irrelevant to philosophy what actually happens in the world. If this were not so how would philosophy differ from the sciences whose business is with facts? It has been suggested, evidently in order to leave a place for a philosophy concerned with facts, that somehow or other the facts of science require to be interpreted and that "everyone may claim the right to draw his own conclusions from the facts presented by modern science". As if to remove all possibility of doubt the same writer also states: "This chapter merely contains the interpretations which I...feel inclined to place on the scientific facts and hypotheses discussed in the main part of the book."* We may ask what 'conclusions' can be legitimately drawn from the statement of physical fact that are not part of physics and are not or will not be drawn by physicists, and in what sense can it be said that a statement of fact requires interpretation? No one who understands the meaning of the sentence 'I had bacon and egg for breakfast this morning' feels the need of interpreting it. Yet physical facts are described in this way. Once the sentence is clearly understood no further statement is required to interpret it. The popular exposition of complicated physical and mathematical ideas is necessarily incomplete and leaves the reader bewildered whenever the explanation does not make the matter clear to him in language which he understands. It is this bewildered reader who is invited to make his own interpretation of the facts presented by modern science and to make an emotional reaction serve in place of the intellectual process of arranging his thoughts in clear order.

Whoever has grasped the point of view of the above criticism will see how it is a mistake to think of philosophy as a possible superstructure raised on the sciences. As an example of this 'super-science' theory of philosophy we may quote the following statement from the article on 'Philosophy' in the *Encyclopaedia Britannica* (13), vol. XXI, p. 41:

> Philosophy claims to be the science of the whole; but if we get the knowledge of the parts from the different sciences, what is there left for philosophy to tell us? To this it is sufficient to answer generally that the synthesis of the parts is something more than that detailed knowledge of the parts in separation which is gained by the man of science. It is with the ultimate synthesis that philosophy concerns itself; it has to show that the subject-matter which we are all dealing with in detail, really is *a* whole consisting of articulated members....The sciences may be said to furnish philosophy with its matter, but philosophical criticism reacts upon the matter thus furnished and transforms it. Such transformation is inevitable, for the parts only exist and can only be fully, i.e. truly, known in

* Sir James Jeans, *The Mysterious Universe.*

their relation to the whole. A pure specialist, if such a being were possible, would be merely an instrument whose results had to be co-ordinated and used by others. Now though a pure specialist may be an abstraction of the mind, the tendency of specialists in any department naturally is to lose sight of the whole in attention to the particular categories or modes of nature's working which happen to be exemplified, and fruitfully applied in their own sphere of investigation; and in proportion as this is the case it becomes necessary for their theories to be co-ordinated with the results of other inquirers and set, as it were, in the light of the whole. This task in the broadest sense is undertaken by philosophy, for the philosopher is essentially what Plato, in a happy moment, styled him, συνοπτικός, the man who takes a 'synoptic' or comprehensive view of the universe as a whole. The aim of philosophy (whether fully attainable or not) is to exhibit the universe as a rational system in the harmony of all its parts; and accordingly the philosopher refuses to consider the parts out of relation to the whole whose parts they are. Philosophy corrects in this way the abstractions which are inevitably made by the scientific specialist, and may claim therefore, to be the only 'concrete' science, that is to say, the only science which takes account of all the elements in the problem, and the only science whose results can claim to be true in more than a provisional sense.

First, as philosophy, according to this view, is to be based on scientific descriptions of the world, it is to be based on theories. Any of these is liable to be discarded in favour of another whenever new facts require the change. This means that such philosophy would decide nothing, for it is the scientists who will put forward the theory and decide whether it is convenient to adopt it. In the second place—and this really excludes the need for making the preceding criticism—whatever we say or write is *on the same level* as a statement of scientific fact or it is nonsense. For if we add to the statement of fact, what justification can be made of what we add—as opposed to something else we might have added but did not? In science we pass from facts and theories to other theories by the process of induction and justify the step either by comparing the new theories with new facts, or by deciding that, for reasons of economy of thought, the new theories are more convenient. But a philosopher would have to offer a different kind of justification. He is not occupied with practical affairs. Consequently he cannot say 'if I do *so-and-so* then *such-and-such* would happen and of course I do not want that', or 'you see I wanted *so-and-so*, therefore I had to do *this*', which are examples of the forms of speech in which one justifies actions and decisions in everyday life. Whatever logical relation the philosopher asserts of the statement of fact and what he adds to it, he cannot escape the need to justify his use of the particular system of ideas in which, indeed, the logical connection in question is already set up. And to justify the choice of any particular system, he has no means which is not irrelevant to the purpose on hand and dependent on the accidents of his actual experience. This is what is meant by stating that everything that language permits us to say is on one level of admissibility.

Just as the superstructure theory of philosophy is untenable so is the notion that science is based on philosophy. It is true that science is based on a theory of a very general kind, but this theory is used for practical reasons and not on account of its plausibility to philosophers. In spite of recent attempts to set up idealism in place of the matter-of-fact view that there is an external world of things which we know through our senses it cannot be said that the materialistic conception has ceased to permeate the manner of thought of scientists. By its insistence on the atomic nature of facts, materialism provided a point of view for scientists which enabled them to restrict themselves to definite parts of experience which could be studied in sufficient detail to develop thoroughly tested theories. In spite of this, philosophers still argue in the following way (*Encyclopaedia Britannica, loc. cit.*):

> For it is evident from what has been said that the way in which we commonly speak of 'facts' is calculated to convey a false impression. The world is not a collection of individual facts existing side by side and capable of being known separately. A fact is nothing except in its relations to other facts; and as these relations are multiplied in the progress of knowledge, the nature of the so-called fact is indefinitely modified. Moreover, every statement of

fact involves certain general notions and theories so that the 'facts' of the separate sciences cannot be stated except in terms of the conceptions or hypotheses which are assumed by the particular science. Thus mathematics assumes space as an existent infinite without investigating in what sense the existence or the infinity of this *Unding*, as Kant called it, can be asserted. In the same way, physics may be said to assume the notion of material atoms and forces. These and similar assumptions are ultimate presuppositions or working hypotheses of the sciences themselves. But it is the office of philosophy, as a theory of knowledge, to submit such conceptions to a critical analysis, with a view to discover how far they can be *thought out* or how far when this is done, they refute themselves and call for a different form of statement, if they are to be taken as a statement of the ultimate nature of the real. The first statement may frequently turn out to have been merely provisionally or relatively true; it is then superseded by, or rather inevitably merges itself in, a less abstract account. In this the same 'facts' appear differently because no longer separately from other aspects that belong to the full reality of the known world. There is no such thing, we have said, as an individual fact; and the nature of any fact is not fully known unless we know it in all its relations to the system of the universe, or in Spinoza's phrase *sub specie aeternitatis*. In strictness there is but one *res completa* or concrete fact, and it is the business of philosophy as the science of the whole, to expound the chief relations that constitute its complex nature.

It would be a profitable philosophical exercise to examine what meaning must be given to the word 'fact' in each of its appearances in the above passage, in order to elucidate the confusion in which it stands at present. Many philosophers have attacked materialism on metaphysical grounds, but none of those who have done so has succeeded in producing a theory (or even admitting that what must replace materialism is a theory) which working scientists could be induced to use in thinking of their subject. The failure of all these attempts is to be traced to the fact that philosophers wanted to settle something that could not be settled in the way they adopted.

The business of philosophers has always appeared to be to settle questions of wide variety, but only in recent times has attention been directed to settling the very obviously philosophical question 'What matters *can* be decided by philosophers?' This question refers to an entirely different possibility from that which is implied when I say 'I can lift 150 lb.', the truth of which is to be decided by experiment. The impossibility of philosophers deciding certain questions must not derive from the limited training, skill or inventiveness of philosophers. If it did so, it would be merely stupid to pretend to settle what is or is not possible for philosophers.

When a mathematician asserts that a circle can be drawn to pass through any three points, which are not collinear, he does not mean that a sufficiently skilled draftsman can, after a number of trials, draw a line with compasses to pass through three given dots, nor does he provide a standard of exactness which must be attained in the drawing in order to establish the possibility he has asserted. The circle and the points mentioned by the mathematician belong to the system of ordinary geometrical ideas. What the proposition asserts is a rule of logical grammar about the words 'circle' and 'point' in this system, and no experimental verification is expected for it. So when we examine the question 'what matters can be decided by philosophy', having in mind that no experimental test is to be proposed as a means of settling it, we are confronted with the decisions made by mathematicians, and we see that the rules of grammar which apply to 'settle', 'decide', 'prove' are the same for mathematics and philosophy. In the experimental sciences there appears also the other use of the word and there is often confusion of the two words—'settle' (in the mathematician's sense) and 'settle' (by an experiment).

So it seems that in philosophy one has only to recognise decisions that are already made, just as in geometry one is engaged in exhibiting the properties of geometrical figures. Where does the philosopher find these decisions already made for him? It was thought at one time that they were to be found in psychological examination of his own experience, and to-day we inherit from this point of view the various psychological theories of knowledge and in particular the psychological theory of meaning. It is remarkable that the part played by language in these investigations should have been nearly lost sight of, and it is tempting to form a historical theory to account for this. In doing so one would merely wander from the point that is philosophically important, namely that in this connection language cannot be treated as the subject-matter of a science. Philological theories of language are irrelevant to the philosopher in the same sense that psychology is irrelevant to the mathematician.

The use of language is understood by everyone; it is used to describe what has happened, is happening, will happen, *or not*, to express our desires, our fears and so on. What we say, or write, or signify in any other way, *can* be understood by others (although it may never be!) provided that we use language correctly. The 'correctly' referred to is not that of the linguist, for the ungrammatical sentence is often as intelligible as its grammatical brother; we all know how to translate from 'bad' English to 'good' English. But there is no translation of nonsense into sense. In making sense, therefore, must lie the essence of language. However, this is not the characteristic of the spoken or written word or sign, for the sounds being clearly heard or the visible marks being distinctly seen does not ensure that the listener or reader understands. What else is necessary? The listener or reader must know the system to which the words or signs belong in their present use. He must see the marks and hear the sounds as belonging to a symbolic system, that is, to a language. And if he is to understand correctly, this must be the same language or system used by the speaker or writer.

One of the classical conundrums of philosophy is: "I experience only my world, how does it come about that there is community of knowledge?" The answer is that community of knowledge is possible because of language in the sense just indicated. For its use, language does not depend on the accidents of peculiarity of speech or handwriting, although, of course, these are always present, but it functions in virtue of the logical structure of its parts. Whether the signs are written or printed or whether the equations are read out, one understands

$$(x+a)^2 = x^2 + 2ax + a^2$$
$$(x+a)^3 = x^3 + 3a^2x + 3ax^2 + a^3$$
$$\cdots\cdots\cdots\cdots\cdots\cdots\cdots\cdots\cdots\cdots$$
$$(x+a)^n = x^n + nax^{n-1} + \frac{n(n-1)}{1.2}a^2x^{n-2} + \ldots + a^n$$

as examples of the Binomial Theorem. What is common to these equations, when understood, is the law of binomial expansion, and the test of one's understanding would be that one could write down other examples. Now, all that is required of one in writing another example is to be able to substitute some other sign for the exponent '*n*' (and if greater variety of example is wanted, to substitute new signs for *a* and *x*) provided that the new signs adopted are the ordinary numerical ones or stand for them. One might even leave a blank space in the formula in every place where '*a*' occurs, but one would have to distinguish by the use of brackets or by order of writing the blank space belonging to *a* from that, for instance, belonging to *x*: the brackets then become the signs, of course. What is preserved when we substitute one set of signs for another in the formula is mathematical form, and is shown by the arrangement of the signs on paper, taking into account whatever alterations of arrangement are permitted by the laws of algebra. These two aspects of mathematical form may be used to suggest to the reader what is meant by the logical structure of the parts of language and by the logical form of a proposition.

We have just pointed out how examples of the binomial theorem are made by substituting a different sign for *n*, and have drawn attention to the necessary restriction that this new sign must be an ordinary numerical one or must stand for one (in accordance with some arrangement made regarding the use of the sign). For instance, it would not do to write the word 'orange' in place of '*n*' because 'orange -1', for example, has no meaning unless we have arranged a definition that wherever 'orange' occurs in a formula we are to write, say, '6'. So it is clear that the mathematical form itself determines what signs we are to be allowed to use. The form leaves a place for the sign '*n*', and although we cannot exhaust by enumeration the possible values which may be given to *n*, we know the definite rules for giving numbers which can be used. We shall refer to a sign in connection with the possibility of its use in a particular system as a *symbol.** Thus the word 'red' as used in ordinary language is a symbol belonging to the system 'red', 'blue', 'yellow', etc. When we say 'There is a red cow in the garden', we mean 'red' as opposed to 'blue' or 'green' or 'yellow', not as opposed to 'oblong' or 'lame' or 'blind'. The *complete symbol*—the proposition just stated—has a certain logical form which we bring to light in the same way as we did in the case of the mathematical form, namely, by asking what signs we may substitute in the propositional sign and still preserve the logical form.†

In the case of the sign 'red' proper substitutions would be 'green', 'yellow', 'blue', but not 'lame', 'blind', or 'oblong', even although propositions in which these words are used in place of 'red' may be true on the occasion when the statement is made.

We have just considered a simple example to illustrate what is meant by logical form and to point out that the test which decides whether or not two words *A* and *B* are of the same logical family is that *A* may be substituted for *B* but not *A* and *B* for *B*.‡ In the example we have used, 'There is a red cow in the garden', we might discuss in a similar fashion the other parts of the sentence and become involved very soon in the problem of meaning. We shall not pursue that investigation now. It is obviously possible to express the thought about the cow in the garden with different words arranged in a new order. This possibility of expression bears

* "In order to recognise the symbol in the sign we must consider the significant use." (*T.L.P.* 3. 326.)

† "The sign determines the logical form only together with its logical syntactic application." (*T.L.P.* 3. 327.)

‡ This statement is correct, but its application to the example used may be confused by the possibility of a pattern of different colours in space. The reader will resolve this for himself however.

an analogy to the possibility of expressing the same mathematical form with a new arrangement of signs obtained by transformation according to the laws of algebra.

It is a commonplace that one can express the same thought in words different from those in which it was first expressed, and it is recognised that the same thought may be expressed by means of another kind of sign—for instance by a picture or a model. How can it be said that it is one and the same thought that is expressed in these different ways? We recognise the identity of thought by this: either that the description of reality by the two or more different symbols would be made true or false by the same facts, or that we can translate from one set of signs to the other, making a calculation as one does in algebra. In order that there may be the possibility of unambiguous translation from one set (x) of signs to another (y), to every distinct sign of x there must be only one equivalent expression in terms of y and *vice versa*. The two sets of signs are then two different ways of naming places in the 'logical space' of what is, from the logical point of view, one system. The analogy which induced the use of the expression 'logical space' is with the use of coordinates in geometry. These are used to name or mark places in space (or it may be on a line or surface) relative to a coordinate system of reference. A change of the coordinate system will give new names to the places by assigning different coordinates: and rules for the translation of names from one system to the other can be given depending on the geometrical properties of the space and system of reference. But, of course, the analogy breaks down in this respect, that a logical space is not necessarily a space in any geometrical sense.*

Physicists will most easily grasp what is intended by 'logical structure' in terms of their own subject. That theories which comprise the same possible phenomena have the same logical structure was clearly seen by Hertz when he wrote:

To the question 'What is Maxwell's Theory?' I know of no

* "The geometrical and the logical place agree in that each is the possibility of an existence." (*T.L.P.* 3. 411.)

shorter or more definite answer than the following: Maxwell's theory is Maxwell's system of equations. Every theory which leads to the same system of equations, and therefore comprises the same possible phenomena, I would consider as being a form or special case of Maxwell's theory; every theory which leads to different equations, and therefore to different possible phenomena, is a different theory. *Electric Waves*, Introduction, p. 21.

One theory may have some of its logical structure in common with that of another. This shows itself in the appearance of exactly the same mathematical equations in the two theories. For instance, what is common to the motion of a pendulum bob, and the variation with time of the electric potential difference set up by an electric oscillator, is expressed by one and the same formula, namely, that for simple harmonic motion. The modern use of electrical models to solve acoustical problems shows the same logical structure in these parts of acoustical and electrical theory.

Words may be used quite correctly from the point of view of literary grammar, and yet make nonsense from the logical point of view. To make clear such logically ungrammatical use of words is the business of philosophy. Elucidation is also required of the common use of incomplete sentences. For example, speaking on board ship, I say 'That other ship is moving at 10 knots', and at once the question arises—'moving relative to what?' Unless there is some previous understanding about the system of reference, my statement is ambiguous and its correctness cannot be tested. Another source of philosophical problems is the following. A word sometimes has different logical parts to play depending on context: this leads to an unintentional grammatical pun in which the grammar of one use of the word is transferred to a context in which nonsense results. We have already discussed an example of this in examining what questions can be decided by philosophy. Of a much more elusive type are the difficulties which come from misleading analogies in grammar. In the treatment of questions such as 'What is time?' it has to make us first answer the simpler one, 'How do we measure time?' and then analyse the source of our discomfort when faced with the former question. Philosophy is constantly dealing in dissatisfaction with linguistic notation, and one of its most difficult tasks is to make men recognise that in the notation lies the source of their intellectual discomfort, not in the apparent problem used to express the dissatisfaction. Philosophers have to unravel intellectual knots of this sort. Once a knot has been untied there is no longer a puzzle, and the philosopher may remove his attention elsewhere. But often in unravelling a knot one does not know where to begin, and has to proceed by gradual loosening of the tangle until it becomes clear what is the correct order of the trivial operations which will undo the knot.

The primary concern of philosophy is with the logic of ordinary language. The results of scientific investigation are not necessary to the philosopher in this task, although it sometimes happens that the emergence of a new scientific theory directs the attention of philosophers temporarily to a particular part of language which has caused trouble to scientists either in the course of developing the theory or in teaching it. On the other hand, some philosophers are interested in problems which have arisen in the various branches of scientific knowledge proper. In order to understand what the problems are the philosopher must submit to the technical discipline of the particular science; thus he becomes a professional worker in mathematics, or physics, or biology, and he philosophises whenever he is concerned with logical puzzles. Perhaps the most important function of the philosopher in science is to remove the difficulties of the learner, either of entirely new theory or of the older parts of the subject which often present so many puzzles to teacher and student. In this connection he can do good work by discrediting ingenious but unnecessary theories which are designed by their authors to surmount contradictions and serve merely to hide confusion of thought by their very cleverness. His criticism has value whether the theories in question are of recent invention or have become accepted and are uncritically assumed in the textbooks.

If the philosopher can be said to have a technique he finds

it in the study of symbolism, that is, of logic. In mathematics he sees a logical method which can be used to infer from propositions which do not belong to mathematics to others which equally do not belong to mathematics. The so-called propositions of mathematics which we use in this way show the logic of the system to which belongs part of the logical structure of the non-mathematical propositions. It is the business of mathematicians to investigate symbolic systems and to invent new symbolism. The business of physicists and other experimental scientists is to use language to describe the world. Theoretical invention in physics consists in making new language for this purpose and in applying new symbolism invented by mathematicians. The making of new language consists not merely in the introduction of new technical terms but in the erection of new logical structure, that is, in making a *system* of new ideas or devising a new method of representation.

Why is the invention of new language necessary? In order to answer this question we have to examine what is required of a piece of symbolism. For example, consider the signal on a railway line: there are two possibilities for the signal arm—down or up—to which correspond the two possibilities for the train—pass or stop. (As a matter of fact the system used by the railways is somewhat more complicated than this when warning signals are taken into account, but we shall think now only of the simple system and disregard also the facetious suggestion that the train should go in the opposite direction when the signal is against it.) We see here that the possibilities of expression by means of the signal correspond one-to-one with the possibilities of fact. Although the signals are used to *prescribe* the behaviour of the train they could also be employed in the same way to *describe* what has happened. The *system* of signals 'down : up' is a piece of language; it has logical structure in common with the facts which it is to describe. In physics when we meet new facts—which are necessarily describable in ordinary language—we look for a theory which retains the possibilities allowed by ordinary language that are in agreement with fact, but rules out those possibilities of ordinary language which are not in agreement with fact. That is, what could be described by ordinary language and was in disagreement with fact *cannot* be described by the theory. We may say that the logical structure given by the theory is part of the logical structure of the facts.

We cannot say of ordinary language that *this* and *that* are not possibilities unless *this* and *that* are possibilities allowed by some other logical system, and we employ this other language to describe them. It is the nature of logic that the introduction of this second language is unnecessary, for we show the logic of ordinary language in using it, without any reference outside itself. Logical structure is internal to and self-contained by language.

The very fact that different modes of representation contain what is substantially the same thing, renders the proper understanding of any one of them all the more difficult. Ideas and conceptions which are akin and yet different may be symbolised in the same way in the different modes of representation. Hence for a proper comprehension of any one of these, the first essential is that we should endeavour to understand each representation by itself without introducing into it the ideas which belong to another. Hertz, *Electric Waves*, Introduction, p. 21.

For example, the logic of Euclid's geometry gives the possibility of describing objects in ordinary space and is part of the logic of ordinary language. It is in everyday use by men who have never heard of non-Euclidean and higher dimensional geometries, and who therefore cannot see the possibility of geometrical relationships which properly belong to these other systems. The propositions which assert these relationships in terms which have a meaning in Euclid's geometry will appear to such men to be nonsense. For instance, it is nonsense to assert of ordinary space that if one prolongs a straight line indefinitely one will return to the starting point.

In this way logic limits the possibilities of description: the boundary of language is not between 'what can' and 'what cannot be thought', indeed, to speak of the boundary of language is to break the grammatical rules for the word 'boundary'.

A boundary separates two parts of space, whereas there is no logical place in language for nonsense. If we wished a spatial analogy to represent a logical space we should conceive of the logical space as the whole of a space: there is then no boundary, for there is no space outside. We see, therefore, that the expression 'what cannot be thought' has no sense, for language does not allow a place for it in the way that there is a place for 'what cannot be done' (physical impossibility) which can be described.

Invention and discovery need not come to an end, and scientists have had to learn caution in stating what can or cannot happen. Because of this and probably also because philosophers have frequently made mistakes, some men, when they write on philosophical questions, take on the air of openmindedness and caution that is proper to science, and with false humility find ultimate asylum in the 'unknowable' and 'what cannot be thought'. It is quite true that caution is required of a philosopher, but to be open-minded about a matter of logic is merely not to know clearly what one is talking or writing about, and as there is a place for healthy intolerance of nonsense in all serious occupations, so there is no more appropriate place for it than opposed to open-mindedness assumed in the face of logic.

In contrast with the habitual caution of experimental scientists is the bold attempt by some writers to claim from philosophy an authority and sanction which their theories would not have if they were presented as hypotheses are usually put forward in physics. As an example let us consider the statement made a few years ago that Whitehead's philosophy is generally accepted, at least in general idea, by those who have taken the trouble to study it.* The writer of this statement attempted to base a physical theory on Whitehead's writings, and instead of waiting for the verdict of experience on the usefulness of the theory insinuated its essential correctness *a priori*. Now if Whitehead has produced a theory which is of use in physics, that theory cannot be part of philosophy since a theory depends on what is the

* W. Band, *Phys. Rev.* XXXVII, 1164 (1931).

case in order to decide whether the theory is a good one or not, and philosophy is not concerned with what is the case or what is useful. If this were the concern of philosophy, then clearly philosophers would be engaged in doing, usually not well, what physicists and other scientists are doing. If Whitehead's theory is to be used in physics then it must be on the same footing as any other theory, and the fact that Whitehead has written *The Principles of Natural Knowledge*, etc., is irrelevant to the consideration of it. Of course, the error which we have been considering does not lie in thinking that philosophy decides something, but in believing that it can give authority to any scientific theory. Authority in philosophy is the authority of clear statement as opposed to muddled statement, of the clear thought as opposed to the intellectual puzzle. But a clear picture of the world may be incorrect.

The function of philosophy in the sense with which we have been dealing is to clear up our understanding of the use of symbolism, and to remove the discomfort sometimes caused by our notation. The meaning of 'the philosophy of physics' ought to be 'clearing up understanding of physics'.

CHAPTER II

LOGIC AND PSYCHOLOGY AND PHYSICS

ALTHOUGH in the title of this essay psychology and logic and physics are conjoined, the main thesis in the essay itself is the disjunction of psychology from logic and physics. It is not our business to examine the aims and methods of the experimental science; but because in the past introspective psychology has been so mixed up with philosophical inquiry that psychological theories have been propounded to do the work of philosophical elucidation, we must give attention to the attempts that have been made in theory to divest thought of its logical characteristics. It is a good rule in logic that whenever one finds oneself involved in psychological inquiry, one has missed the logical point; and though to embark on this discussion may seem to break the rule, it will be seen that we follow it. Our concern is the logical one of showing *how* the psychological theories are irrelevant; that they must be irrelevant has already been maintained.

The study of the process of learning is of great importance to psychologists and to physiologists interested in the central nervous system of man and animals. In this study the theories of the association of ideas and of the conditioned reflex have been applied. It is not for us to examine here how well these theories describe the facts of the experimental sciences referred to. Our task is to show their irrelevance for logic. In learning my way in a great city such as London, I am conscious of having to recapitulate the history of my movements in it, but after I have become familiar with the plan of the city, I do not need to make this recapitulation when I wish to go from one place to another in the city. Indeed, most of us, having memory of very limited power, would find it impossible to perform the feat. This is not to deny that there may be some cerebral process which does make some kind of recapitulation, but to assert that irrespective of the decision of experimental science on this matter, the recapitulation is not logically necessary. How many of us can recall the long process by which we learned to count? Yet who of us will assert that we are unable to make arithmetical calculations because of this lost record? Evidently the historical accidents of the learning process have as little to do with knowledge of arithmetic—or of any other matter—as the accidents of handwriting have to do with the sense of the words we write. And it is so with ordinary language. It is not necessary to remember the accidents of how we learned to speak and write and read in order to understand. What we learned is logical structure, and just as a geometer would laugh if it were suggested to him that he can call to mind the proposition he learned about a triangle with the aid of a rather poorly drawn figure labelled *ABC*, only if he first thinks of that particular drawing, so it is difficult for one to regard seriously the contention that the association of ideas by psychological or physiological habit is a substitute for logical connection. Regarded as a matter of psychological fact, particular ideas *might* be associated in a different order from that observed; in logic an idea belongs to a system of ideas and the logical connections in the system *cannot* be otherwise than they are —language does not admit the possibility of exhibiting these ideas with different logical structure. The relationship of ideas in a logical system is an internal one, the psychological association of ideas is by external relations of temporal order and so on, which might be otherwise, for language allows us to say what the relations would be if this were so.

What we have to get clear about is the distinction between logical connection and psychological or physiological causation. Let us consider what is involved in reading from a book or a musical score, and in any copying or translating process. On the causal theory, the pianist who is seated at his instrument with a score in front of him, fingers the notes he does in fact strike, not because he intends to play what the score directs, but because of processes which we shall leave to the ingenuity of scientists to invent. According to this view, the musical score plays the kind of part which a drug or surgical operation does in the behaviour of the patient. One is entitled to ask then how the player is to understand that he has not played the score correctly—indeed, how correct and incorrect playing can have meaning at all. On the causal theory, in order that the player should play something different when confronted with the same score, we have not to appeal to his understanding of music, but to supply some other stimulus which would again act as a drug and so on. Whereas in actual life what we should say is something like this: 'You have made a mistake; you have played E here' (pointing to the keyboard) 'whereas the score has F there' (pointing to the place in the score). In fact the score is played by means of a particular rule of projection or translation from the score to the keyboard. When the rule is followed the score is read correctly; when it is not, it is read incorrectly. By considering the process of copying where it is necessary to specify the rule explicitly, mentioning size and so on (whereas in reading, the rule is always taken for granted, and its existence therefore overlooked), we can bring out the part played by the definite rule without fear of misunderstanding. If we make use of certain rules of projection to copy a circle we can obtain an ellipse, and these figures are similar to each other with respect to the rule of projection employed. (To say that two figures are similar without understanding a particular rule of projection, is to abuse the word 'similar'.) When in ordinary language we speak of making an 'exact copy' we mean a copy according to the rule that corresponding lines shall be of the same length and corresponding angles equal. Here again the causal theory is faced with the problem of giving a meaning to correct and incorrect copying. Deny the use of the rule and you abolish the criterion for judging correctness. And without mentioning the rule, how is one to describe the intention to use it, which is none the less definite on account of mistakes made in execution? Clearly one cannot. Whatever theory of experimental science is put forward to describe an intention to act according to a particular rule, the rule in question remains a logical datum, and this rule may be different from any law which the experimental scientists could adopt as adequate to describe the actual action; for action, although intended according to the rule, may not in fact follow it. For example, I write

1	2	3	4	5	6	7
1	4	9	17	25	36	49

intending to use the rule of squaring in passing from the numbers of the first row to those of the second. But the number under 4 is in error. A mathematician could give a wide choice of comparatively simple formulae by which one could transform the numbers of the first row into those actually written in the second, but the law of squares cannot be admitted unless we make another rule that numbers of the second row are allowed to differ by an error of, say, 1 from the value calculated according to the rule of squaring. Making the statement more extreme, we may say that even if none of the numbers of the second row were correctly written according to the rule of squaring, that would not contradict the assertion that the rule of squaring was intended. For performance is not the expression of intention, but the attempt to fulfil it. One might as reasonably confuse the plan of a building with the actual building itself as fail to distinguish intention from action.

In logic the justification of an action or particular behaviour is found in a reference to the intention, plan, or prescription to act in that particular way. For instance, a motorist stops at a traffic signal which is against him, and is asked by some-one unaware of the signal why he has stopped. The driver justifies his action by saying, 'I stopped because the traffic signal indicates "stop"'. Whereas, if an experimental scientist were asked professionally why the motorist has stopped, he would explain that the engine of the car had been thrown idle and the brakes applied as the result of certain levers having been operated by the driver. If further explanation were expected, a description would be given of the nerve and muscle processes which resulted in the pulling of the lever, and finally a theory would be put forward concerning the physiological and psychological effects of the 'stop' signal on the driver of the motor car. But, however ingeniously and successfully this theory may account for the motorist's action from the scientific point of view, it is not a substitute for the logical justification which the motorist himself offers for his action. We must distinguish the 'because' of logical justification from the 'because' of physical causation (i.e. in terms of scientific theories). An example which will help to illustrate the matter further is the following. Suppose that one's calculation in arithmetic is called in question. The justification would proceed by repeating the calculation in detail, and statements such as 'That is correct because $4 \times 3 = 12$' will occur in the exposition. No one would attempt to justify a calculation (whether correct or not) on the basis of a theory which describes the psychological and physiological processes consequent on the appearance of the collection of signs by means of which the problem was stated. Such a theory might explain scientifically why on one occasion the calculation was made correctly and on another incorrectly. Nevertheless, the decision as to when the calculation is correct does not lie with such a theory but with logic, and the justification of a calculation requires reference to the rules of calculation which are logical and not psychological.

The essence of the distinction between the logical justification of the motorist's action on seeing the 'stop' signal and the scientific account of how and why he stopped is that in the former 'stop' must have a meaning and in the latter it need have no meaning at all. So far as the scientific theory is concerned any other sign or event *might* have produced the result of causing the driver to stop his car without even involving the rule of the traffic signs in accordance with which the motorist acted. The motorist gives the rule in his justification, and we know what it means to follow the rule if we understand it.

The invasion by psychological theory of what is clearly a matter of logic rests on the widely spread inclination to recognise only what is perceptible as existent, and to deny and overlook whatever is not evident to the senses. Frege, in his *Grundgesetze der Arithmetik* (1893), showed the error of such an attitude by asking how we are to be satisfied with the numbers of arithmetic which are not sensible objects. The numerical signs are visible, but we cannot, as is sometimes done, assert that the signs are the numbers, for the signs have properties quite different from those of the numbers they stand for. Nor can we, as is sometimes said, impute the desired properties to the signs by means of definitions, thus securing the characteristic property of the number in a sign which is visible and therefore exists. A more subtle answer to Frege's question and one widely adopted by mathematicians is to regard the numerical signs as chess pieces and the so-called definitions as the rules of the game. The signs then represent nothing at all, but are themselves the objects of arithmetic. An arrangement of chess pieces does not represent anything, whereas by $3^2+4^2=5^2$ we express a thought; this is what the 'chess theory' of arithmetic overlooks (Frege).

It has to be made clear what definition is and what can be achieved by means of it. A definition is commonly believed to have creative force, while as a matter of fact nothing results from it but the clear indication of something which is assigned a name. The geographer does not create a sea by drawing its boundary and saying 'the stretch of water bounded by this line I will call "Yellow Sea"', and neither can the mathematician by his definitions create anything intrinsic. It is intellectual superstition to believe that an oval figure drawn with ink on paper will be obtained by definition of the property 'this added to unity gives unity'; one could just as well make a dirty schoolboy clean by definition. Lack of clearness on this matter, wrote Frege, comes from muddling the distinction between idea and thing. When we say 'A square is a rectangle in which adjacent sides are equal' we are giving the property the figure must have in order to fall under this concept. But the idea 'square' is not a rectangle, just as the idea 'black cloth' is neither black nor a cloth. We define an idea by giving the property an object must have in order to fall under the concept or idea, but this does not ensure that an object will in fact be found with this property.

When we are asked for the explanation of the meaning of a word we give it by means of a definition. The definition given in a dictionary, being mere substitution of another verbal expression for the word we have in mind, seems not to get us any farther, whereas when we give an ostensive definition we are tempted to believe that something final has been reached. For example, suppose that someone does not understand the word 'orange'. We pick up an orange and pointing to it say, '*This* is an orange'. The question arises whether or not this act of pointing at the fruit held in the hand necessarily shows the meaning of the word 'orange'. Clearly no, for the shape of the orange might just as well be intended. What we have done is to give another sign in place of the spoken word, namely, the gesture of pointing at the orange held in the hand, and this gesture could be employed wherever the spoken word is used. The sentence 'Take the orange' could be rendered 'Take —', where '—' stands for the gesture referred to, and so on. The dumb sign or gesture appears to be suggestive in many cases, but logically it is on the same level as any other sign for which it is substituted. We have to learn the correct use of each sign. We do not exhibit the meaning of 'red' merely by pointing at a number of red-coloured objects. For the meaning of 'red' is bound up logically with the meaning of 'blue', 'green' and so on. What we learn when we learn the meaning of these words is a piece of logical grammar. The meaning of words is shown in their application. The word 'square' has not the property squareness, but the rules of grammar for its correct logical use show what geometrical properties the figure referred to possesses.

The meaning of 'red' is a good case to examine, for it is not possible to complicate the matter by means of an ingenious description, and because the psychological theory of meaning seems here to be on its most familiar ground, associating the word 'red', which has nothing coloured red about it, with some image or other of a red object or patch. In disposing of this theory and any other which works by referring the word to an image or other sign, we have to notice first that the red-coloured image might equally well have been produced by sight of the appropriate red-coloured object, and that therefore for our purpose the image 'in the mind' is on exactly the same level as the red-coloured object visible to anyone else. We may speak of objects in place of images. Suppose a man were asked to colour some woodwork yellow, and that he goes to select his paint. According to the image theory, some sign will be required—Wittgenstein humorously suggests the ringing of an electric bell (!)—which will tell the man when he looks at the paints, 'This is yellow paint'. The question then arises, how is he to know what the ringing of the bell means, and so on. Thus the second fault of the theory is that it fails to recognise that so far as meaning 'yellow' is concerned, the yellow-coloured patch or image and the word 'yellow' are on the same footing. To state this is not to suggest that it may not be a psychological fact that the hearing and speaking of the word 'yellow' are accompanied by images of yellow-coloured objects: even if this is experimentally true, it is none the less irrelevant here. For the yellow-coloured object or patch could mean something quite other than yellow. The yellow marks used by cartographers, for instance, do not stand for the colour yellow, and the red of the traffic danger signal does not stand for red.

Whenever we are concerned with the meaning of symbols, we have to deal with the use of signs in a language—whether the signs be written, spoken or gesture. In common usage, the unit of expression for conveying sense is the sentence; the logical category which includes the sentence and also non-verbal units of expression is the proposition.* In using this word, we do not imply the specialised sense of the term which is understood by mathematicians and logicians. A proposition is a picture of reality—descriptive of the past, present, or future, or prescriptive as a command, intention, desire, fear, and so on. If it is descriptive it will be true or false when the reality described has taken place, and if prescriptive, obeyed, fulfilled or not. We understand a proposition when we know how to verify its truth. The method of verification is therefore

* It is interesting to draw attention to the fact that the Latin word *propono* is a translation of 'represent' in the sense 'point out'.

essential to the meaning of the proposition. It is part of the logical structure of the language to which the proposition belongs. The proposition selects one of the possibilities admitted by the language and says 'reality is like this pictured possibility'. On the other hand, the proposition may assert that reality is not like the picture in question. It is evident that we cannot leave out the grammar of 'yes' or 'no' in representing reality. No matter how faithfully we may reproduce the situation we intend to describe, unless we make use of the system 'yes : no' it will be in doubt how our representation is meant. For instance, a route to be avoided is pointed out by drawing it on the map just as the route to be taken. To show the difference we must mark the routes in different ways, but whatever device we use, it has to be learned at some time like other conventions for reading the map. Again, an arrow points in the direction to be followed, not because the arrow suggests the way to go, but because of the convention to use the arrow in that way. The system of direction arrows would work equally well if the arrows were understood in the opposite sense. This is characteristic of any proposition. It must make sense to deny it. Making sense has nothing to do with truth or falsehood. The former is a matter of logic, the truth of a proposition is verified by holding the proposition up to reality. A statement which is qualified in such a way that the test of its truth or falsehood cannot be made is not a proposition.

I have used the word 'reality' and this seems to require some explanation, even although we all know how to use the word 'real'. By 'reality' I understand what actually happens, not in any subtle and intricate sense, but merely as opposed to that which might have happened but did not. The term 'real' has two meanings; on the one hand, as opposed to illusory or hallucinatory, and on the other, real is opposed to the ideal which includes what might have happened and did not. As the result of the confusion of these two distinct uses of the word, attempts have been made to apply the legitimate investigations of psychologists into illusion and hallucination to throw light on the nature of the so-called ideal, and have usually ended by putting, for example, number on the same level as hallucination. The essence of the matter is that logic permeates both real and ideal, and both real and illusory. The predicate 'real' and its opposites get no nearer logic than the predicate 'algebraic' (of a curve). The logical structure of the proposition by which we describe a particular state of affairs is the same whether we are under illusion or not, and, on the other hand, as we have stated above, the fact that the proposition is not true does not affect the logical structure of the proposition. That fact lies outside the proposition; it is an accident, the proposition might have been true.

Only propositions are true or false. The word 'true' stands for the relationship which a proposition may have to the reality it depicts. The rôle of 'true' is like that of the word 'correct' as applied to reading and copying. Here psychology, as a substitute for logic, is in the same predicament with the truth of a proposition that it is with the correctness of a copy. The representing relation by which the proposition is to touch reality is part of the proposition. It is not a question of the proposition resembling what it depicts and then our depending on some psychological process to signal this resemblance to us as in the 'electric bell theory'. We do not require any sign to stand between the words 'true' or 'false' and their application. If we assert the contrary, then 'true' becomes the name for another sign (e.g. the ringing of the electric bell) and we are left with the problem of connecting this new sign with the proposition and the reality it depicts; that is, we have come no nearer to accomplishing what we set out to do, namely, to give a psychological explanation of the meaning of 'true'. In any case, to speak of resemblance between the proposition and reality is to beg the whole question, unless we admit identity of logical structure, which is not a matter of psychology at all.

The essence of this matter of language and logic is that nothing can stand between language and its application. This is not to say that language is simple. It is not. The silent adjustments which are required to understand colloquial language are enormously complicated. These adjustments are all part of language; whatever their interest for psychologists as phenomena, and whatever theories are put forward to account for them, their status in language must remain a matter of logic. It is not the particular adjustment by itself which has linguistic significance, but that it is this particular adjustment opposed to other adjustments of the system to which it belongs.

What, then, of thinking itself? Surely, it will be said, this is a matter for psychology: yet so far as the logical character of thought and the thought process is concerned, we must answer 'no'. For the thought process is the symbolic process, the process of operating with signs; whatever signs, visual or other images, make up the procession—there is no thought without them—it is their connection in systems of notation that makes them significant. Logically we are confronted with the old problem of the pianist playing his score. Is he reading from the score or not? Am I thinking or is it just a jumble of words and pictures? Whatever psychological distinction can be made between these two possibilities on the basis of experimental fact is added to the logical distinction and does not replace it. Unless the logical distinction is already known, how can it be said that the psychological facts relate on the one hand to connected thought and on the other to a series of disconnected images, which express no thought?

From the point of view of logic, thinking is calculating. Whenever we have to plan our affairs we calculate, not merely in those cases where numbers are obviously required, when we use ordinary arithmetic, but every time we represent to ourselves the possibilities for action and select from these according to some rule which we choose to use. The detail of representation is decided for us by our needs. Often the representation is very crude, but we must not therefore be misled into asserting that one cannot calculate by such apparently inarticulate means. For example, I am planning to accompany my friend *A* to the theatre next week. Each of us looks at his diary, and considers each day in turn: Monday is not possible because *A* has some business to attend to; on Tuesday I am invited to Mr and Mrs *B*'s; on Wednesday we are both free and also on Friday, but not on Thursday and Saturday. We have therefore to choose Wednesday or Friday, all the other possibilities have been ruled out. We later decide in favour of Wednesday because we are able to secure good seats only for that evening. We complete our plans by arranging to have dinner together before the performance, at an hour determined by the time of raising the curtain, the interval required for the meal, to get to the theatre and so on. This calculation is not an exact one with respect to times, if it is easy to fit this plan into the rest of our programme for the day, but whenever one is very busy it is necessary to plan with much more detail in order to make possible the fulfilment of the engagement. It is to enable us to map out detailed programmes that reference to the clock is necessary. Very often, when there is plenty of time available, we plan in accordance with what we wish to accomplish. Here there is no reference to the clock in our plan, yet the plan can be quite definite, expressing the intention to perform such-and-such task after this one and before that. The fact that the possibilities from which we choose in planning are crudely conceived compared with those given by a carefully drawn time-table, does not in any way impair the determinate character of the choice when considered relative to the proper system of possibilities. Of course, relative to the time-table, the plan appears vague, and we can ask questions which are legitimate with respect to the time-table but which it is impossible to answer from the plan adopted on the other basis. As another example—when we say that a chess piece, say, a black knight, is on the square KB 3 we have given a perfectly definite piece of information. It means that the knight occupies that particular square as opposed to all other possible squares on the chessboard. But of course this information does not specify the geometrical position of the object nor its orientation with respect to the board. From the point of view of fixing the position of the knight on the board the naming of the square is crude compared with specification in terms of measurements made with even the simplest physical apparatus. On the other hand, so far as the game of chess is concerned, all that is required is to name the square. Naming the square selects one possibility for the position of the piece in question on the chessboard, and rejects all others.

These two illustrations should help to make clear the sense of the statement that thinking is calculating. In a mathematical calculation, one passes from one line to the next by using the rules of calculation to transform the signs of the former into those of the latter. The rules by which we connect the images and words expressing our thought are those of logical grammar. Language, in its general sense, is the mechanism of calculation. Just as a geometrical figure cannot be presented in a model contradicting the laws of the geometry of space, so our thought is governed by logic; but just as the mathematician can make mistakes and assemble signs in meaningless array, so we may make nonsense by the incorrect use of words. A thought about the arrangement of things in ordinary space expressed by means of a three-dimensional model, while it might not agree with the facts, must conform to the logic of space; on the other hand, the expression of the thought in words might correspond to no spatial possibility at all. In this respect there is a distinction between thinking in terms of things and thinking in terms of words. The latter is more easily liable to error, and in the extreme cases when the rules of philological grammar serve in place of logic, or when, as in poetry, words are used not as symbols but as specific drugs to create images and produce emotional response in the reader or hearer, the words can no longer be given their proper meaning, and the ordinary symbolic process is frustrated. Accordingly, such use of words has no further interest for the man who is concerned with language as the vehicle of scientific thought, and although popular works on science and philosophy afford many examples of this abuse, and although the technical journals are not yet free from it, scientific men in their professional affairs look with favour only on the correct application of their language. Any other use of words, that is, where they cannot be given their proper meaning, is matter for study by artists or psychologists.

Let us now consider the serious attempts which have been

proposed to do for physics what Whitehead and Russell have done for mathematics in their *Principia Mathematica*. The proposal in question is to make a logical analysis not merely of the framework which forms the subject-matter of theoretical physics, but also of the operations which the experimenter has to perform in the laboratory whenever a physical fact is observed. The aim of the analysis is to show how physics as an activity is made up from certain elements, 'simples' or 'indefinables', and it is claimed that since such analysis inevitably introduces the psychological experience of the experimenter, so we cannot have a proper philosophy of physics without psychology. In other words, while Whitehead and Russell were concerned with logical elements only, in physics we must also have psychological elements—namely, our elementary sense perceptions. Imagine that such a scheme has been worked out. On the one hand, it is based on a psychological theory of how experimenters work, and on the other it introduces a new terminology to state what, generally speaking, has already been expressed with sufficient clearness to be understood in technical works on experimental physics. Only in one respect can such an analysis be useful to physicists. It may remove some of the indefiniteness of statement in the experimenter's handbooks. In so far as a psychological theory is advanced it is very likely to be an atomic one. It may be a good theory in the opinion of psychologists and neurophysiologists. Even if such a theory has much to recommend its adoption by scientists in general when examining their scientific experience, it is nevertheless certain that the theory is not *necessary* for physics. Language as at present in use is a vehicle adequate to express the experimenter's thoughts. After all, the physical event does not take place in the observer's head, otherwise one might expect that in order to establish what happens in a physical experiment, the observer ought to be under the observation of a brain surgeon, whose observations would in turn have to be subjected to the same test and so on *ad absurdum*. The essence of the physical experiment is that the observation, whatever it is, is described by a physicist (*A*) in language which another physicist (*B*) can understand, and that observations can be made by *B* similar to those made by *A*, and described by him in the same language, although not necessarily describing the same state of affairs. The worries of the solipsist need not concern us here at the moment. This language does not refer to places in *A*'s (or *B*'s) visual field, nor does it speak of *A*'s private kinaesthetic and auricular sensations; it gives places in ordinary space surveyed by measuring rods and reports sounds heard at particular times and so on. Roughly speaking, there is none of the technical language of psychological analysis in the physical laboratory, and, it may be added, neither does laboratory work consist merely in observing pointer-readings.

The psychology of physicists as a professional group engaged in their daily work may be a useful study for psychologists and may even have to be made at some future time in the public interest. Similarly it is possible that biologists may some day study the activity of physicists as a biological phenomenon. But the study of physicists from either of these points of view is external to physics—such studies are not the business of physics any more than psychological and biological studies of the activities of musicians have any logical place in the making of music or in theories of musical form.

The distinction which has just been drawn seems at the present day to require special emphasis. More and more attention is being given to the relation of science to society, and in the process some curious methods have been used in the attempt to make scientists recognise their social responsibilities. One of these is based on the implicit suggestion that biology is a substitute for logic. The 'humanists' both in and out of the ranks of professional science assail the 'purists' by asserting that logical matters are decided by experiment, and that even the result of a calculation is obtained by a process of trial and error. Surely this is to carry antischolasticism too far. The very fact that such assertions are made shows that the authors have in mind an authority other than appeal to experiment. Does one have to make an experiment to discover that a given collection of sounds is a possible intelligible sentence and not an incoherent babble? And what experiment is necessary when one recognises a well-known tune correctly played, or the binomial theorem incorrectly applied? Let us cultivate the habit inculcated in the teaching of elementary mathematics and as a discipline in these matters write down 'Given: the logical form of what we are talking about'.

CHAPTER III

ON METHODS OF REPRESENTATION

THE physicist, like other natural scientists, employs language—both that of the ordinary man and his own special symbolic methods—to represent the world of fact. He does this for purposes which may be clearly evident and explicitly stated, as for example, when the application is an engineering one, or, it may be that these purposes are more closely related to the needs of the pure scientist. So much do we take for granted the use of ordinary language and of its technical counterparts in our professional life, that we are guilty more often than not of overlooking the conscious processes of representation. In everyday affairs a copy of a picture is asked for and supplied without its being necessary to specify in so many words that the copy is to be made by ordinary similarity and not by the rules of some possible distorting process of projection. An artist when asked to paint a portrait has not the liberty to paint the image of his subject as seen in a convex cylindrical mirror. To state in so many words the rule for copying generally appears pedantic. Nevertheless, in the ellipses of common usage is a habit which the physicist may easily take over into the business of representing nature. When he does not have clearly in mind the conventions of his methods, he is likely to be led into misuse of any bad notation which he may have, and ultimately into asking philosophical questions which arise therefrom. It is hardly necessary here to insist on the importance of notation as an aid to thought in science. In the past, however, emphasis has been placed on special notation as an aid to complicated thinking; in contrast, let us consider how a notation may trap us into error. Imagine the orthogonal projection of a sphere on its equatorial plane. The entire spherical surface is represented on a part only of the plane. Points outside of this part of the plane correspond to no real point on the sphere. Some of the questions we ask ourselves about physics are rather like the question, 'what is represented by the point *A*?' where *A* is on the plane but outside the projection of the sphere. The answer is, of course, that *A* represents *no* point of the sphere. The trap is presented to us because the method of representation employed allows us to mark the point *A* and, what is more, appears to allow the same grammatical expression when *A* is substituted for *B*, which does represent a point of the sphere. While we might invent another method of representation which would not present us with this linguistic temptation, it is not necessary, provided that we apply the rule of logical grammar for the correct use of the former method. The instance which we have just inspected seems trivial, for the logical point is so obvious whenever we think of the geometrical relation between the sphere and its projection; yet one does not have to go far to find a parallel in physics which is as direct and appears to cause confusion of thought.

Think of the spatial representation of time. The succession of points along a line only in one direction corresponds to the succession of instants of time. The motion of a particle along a straight line with uniform velocity can be represented by the equation $x=vt$, that is, by a straight line in xt-space. In the motion, this line is traversed in the direction away from the origin of time. We are tempted to say that the same line considered in opposite sense (i.e. towards the origin of time) represents the 'motion of the particle backwards in time' instead of saying that the line viewed in this way does not represent a process at all. The logic of 'before : after' does not allow sense to the form of words 'backwards in time' any more than the geometry of the sphere allows its orthogonal projection on a plane to cover the whole plane. Nevertheless, literary grammar (as opposed to logic) appears to sanction the expression 'backwards in time'. There is no backwards in time, only backwards in space. The analogy we consider is something like this—a history of events can be written in a book, and we may turn back in the book to the record of more distant past events. The very fact that in speaking of temporal relations we use words whose application properly belongs to space, ought to warn us that when we have these terms in mind, we use a spatial representation of time. Recognition of such a simple matter prevents the further temptation to theorise about entropy and 'time's arrow'.

How often do we use the figure 'time like a river'! The measurement of time means that we compare with another process those processes whose temporal course is to be described. In physics, of course, the process of reference is standardised; but in ordinary language we do not necessarily refer to the clock, we leave the reference process unspecified, that is, a variable like x in algebra, and any process will serve as a particular value of the variable. The figure of the river is a misleading one because it suggests events strung together on a line like the pictures on a cinema film, which is pulled past us, and that if only we could arrest time we could see past, present and future together, just as we can see the river or film spread out before us. The same idea is inherent in much of our thought about any process. Take, for instance, a musical tune. It is as if a gramophone record of the whole tune were in course of being played. It is all 'there',* but we hear only a little bit at a time. This same analogy pervades nearly all epistemological theory and it is nearly always wrongly applied.

In using analogy we must have caution. If the method of representation which we denote by A is analogous to the method B, it is necessary to avoid the temptation to employ the grammatical forms which belong to A when we are speaking of the representation by B. Roughly speaking, the word 'analogy' has a bad reputation in science and particularly so among mathematicians. This is perhaps surprising to anyone who considers the growth of physics. The theory of electrostatics borrowed some of its symbolism from elasticity, and optics borrowed from elastic solid theory. More recently wave mechanics was developed from classical mechanics in terms of the analogy between mechanics and optics. Whatever the newest views of quantum mechanics may be, it cannot be

* One is justified in asking, where?

denied that this analogy played a useful part. To-day, it may almost be said that the mathematicians wish to suppress the analogy or at least not to emphasise its existence and historical place in the subject. Is there any explanation of this state of affairs? Let us look at two analogies in the history of physics and we shall find something to guide us. First recall to mind the model of the aether devised during last century to explain Maxwell's theory of light. This analogy usually raises the question of the existence of the mechanism which is postulated in the explanation. As a matter of fact it need not raise this question, as we can see when we think of electrical models of acoustical systems and recognise that there is no question of the existence of coils of wire or of condensers in the acoustical apparatus. It is noteworthy that we should accept easily this electrical-acoustical analogy without being tempted to ask philosophical questions, whereas with the theory of light in terms of a mechanical model, philosophical questions have always been to hand. We recognise clearly in acoustics that the analogy in question is completely expressed by mathematical equations, but we also find it convenient to use some of the electrical terms when we wish to name quantities which have the same mathematical properties in the two theories. In electromagnetic optics, however, the mechanism of the model of the aether does not help in this way at all. Once we have expressed by means of Maxwell's equations the electromagnetic connections between different places at different times, we have an adequate symbolism for representing optical phenomena. We no longer feel the need even for a medium of which to predicate undulation. The equation $x = a \sin nt$ represents the vibration of a particle or the oscillation of an electric current and so on, but the equation does not vibrate and neither does a graph of this function. (We could, on the other hand, represent the oscillating electric current by means of a vibrating particle such as a pendulum.)

The mathematician's objection to analogy arises in part from this: that the mathematical equations which express the law common to the method A and its analogue B are all that is essential to the analogy. One does not require the picturesque and in some respects misleading language appropriate to the analogy. Such forms of speech can be dispensed with, for correct application of mathematics does all that is necessary. We find therefore that although analogy may serve useful purposes in the development of theoretical ideas, in the course of time reference to the particular analogy almost inevitably drops out of the subject because one comes to recognise the immediate applicability of the mathematical theory which was in the first place used via analogy, and secondly because, generally speaking, the analogy breaks down and becomes therefore a hindrance to correct thinking about the subject to which it had been applied. The analogy itself never shows that it is misapplied; this is shown only when the logic of the analogy is compared with the logic of the possibilities it is used to describe, and such comparison is usually a tedious and difficult process.

The well-founded attitude of caution in the use of analogy has tended to become transformed in some quarters into the attitude of objecting to any use whatever of analogy. When Heaviside invented what has now become the operational calculus, mathematicians were at first unwilling to admit the validity of his methods. This calculus is based on the idea that if p denotes the sign d/dt of differentiation, then p can be treated like an ordinary number in algebra. Any analogy of this type, according to which one proposes to apply to a sign used in one calculus (C) the rules for calculation in another calculus (C'), is an interesting one, and as a rule the mathematician as a critic of innovation is liable to occupy a strong position. It is quite likely that certain combinations of signs, although permitted by the rules of C', have no meaning when applied to C, and that ambiguities arise as to the application of the rules of the calculus C to the result of a calculation by means of C'. For example, suppose that in C the signs do not obey the commutative law of multiplication, whereas in C' they do obey that law, then there will be a doubt as to how the result of a calculation by means of C' is to be read whenever it involves a product. One should observe, however, that so long as one avoids such ambiguities and meaningless combinations of signs, one may be able to use the analogy in question quite effectively in applying mathematics to physical problems. For instance, without introducing a student of physics to the operational calculus as a body of mathematical knowledge from which the last trace of ambiguity and lack of clarity has been removed, one may teach him to deal with alternating current circuits by means of a generalisation of Ohm's law which is in effect to treat Kirchhoff's laws by operational methods. As a result the student learns a symbolic method which is very powerful. Until he has to apply the method to problems which refer to the subtleties of the operational calculus, however, he is not *compelled* to know more about the calculus. Now this principle applies not only to mathematical but also to physical representation in these very same problems. The so-called residuals in his resistance boxes and impurities in his condensers and mutual inductances are not taken into account until the experimental results call for such interpretations.

Under the influence, perhaps, of our mathematical mentors, we are often too ready to regard physics as a single theory (relic of philosophical monism) that should be presented like Euclid's geometry, as if it were one method of representation which, possibly owing to the intellectual laziness of experimental physicists, never appears to be presented as a logical whole, although it could be. In the writings of mathematicians on physics, however, one seems to see the subject unfold itself with a certain inevitability hardly, if ever, referring to the laboratory or to the experimental facts which support the structure. As if one could say *a priori* what the world must be like. Now physics is actually a conglomeration of methods of representation. Two methods of representing the same phenomena must agree in some way, otherwise one method necessarily gives a picture wrong in every respect when the other gives a correct one. What is common to the two pictures by the different methods of representing is like what is common to two photographs of the same scene printed from half-tone blocks with differing screen mesh. If the meshes differ greatly in size, one photograph will be a picture with cruder detail than the other, but so long as the mesh is not made so large as to prevent the possibility of one's recognising some of the form of the picture, one is still able to agree that they are pictures of the same scene by different methods of representation.

The cheerful dismissal of Newtonian mechanics by relativists ignores just this matter. It is not a question of Newton's laws being wrong and Einstein's being right. In many cases our experiments do not yield pictures of grain fine enough to represent what the relativist asserts should be the case. For such pictures Newton's laws are quite exact, and naturally, on account of the greater simplicity, we employ this system in calculating what is to be expected. Only when the experiments call for relativistic representation do we employ that method. The fact is that in physics we choose the particular method of representation adequate to the purpose in mind, just as a carpenter chooses on one occasion a saw and on another a plane to give a particular shape to a piece of wood. In electrical practice one sees a like multiplicity of method. In a circuit to be used with direct current, one does not take into account its inductive or capacitive properties unless one is concerned with the effects of transients on the line. Even when one is concerned with transients one does not apply the whole Maxwellian theory which requires us to know the geometrical arrangement of the conductors and insulators in space—we make a schematic diagram of the circuit inserting resistance, inductance and capacity and apply Kirchhoff's laws. Only when the time scale of the process is sufficiently brief do we have to give up quasi-static conceptions of the electrical properties of the network and describe the entire electromagnetic process in space by means of Maxwell's equations—assuming that sufficient mathematical skill is available to accomplish the calculation.

When one has to deal with atomic phenomena in crude fashion, one uses the classical approximation to quantum mechanics to make the calculation. There is then a certain analogy between quantum mechanics and classical mechanics.

This analogy is worthy of study for it is exhibited by the symbolism explicitly. In classical mechanics, for a system whose Hamiltonian function is $H(p, q)$ we write the Hamilton-Jacobi equation

$$H\left(\frac{\partial W}{\partial q}, q\right) + \frac{\partial W}{\partial t} = 0,$$

whereas in quantum mechanics we deal with the system whose Hamiltonian function has the same form by writing

$$\left[H\left(\frac{h}{2\pi i}\cdot\frac{\partial}{\partial q}, q\right) + \frac{h}{2\pi i}\cdot\frac{\partial}{\partial t}\right] W = 0.$$

These two equations have the same form, if in the former we write p for $\frac{\partial W}{\partial q}$ and in the latter p for $\frac{h}{2\pi i}\cdot\frac{\partial}{\partial q}$ etc. This identity of form is the analogy between the two methods of treating the system defined by $H(p, q)$. We expect, therefore, and it is so found, that the two methods give the same result to calculation of the motion provided that a crude picture only is in question. That does not mean to say that classical mechanics has the possibility of presenting internal atomic motions any more than geometrical optics has the possibility of predicting the diffraction of light. What is important for us is to recognise that the identity of form to which we have already referred insures identity of the possible pictures of fact by the two methods under the proper circumstances as to the scale of the phenomenon. Analogies of this type which are exhibited in mathematical form are always instructive and help one to bridge the gap in learning any new theory which is intended to work where another theory, otherwise successful, has failed; and such an analogy will give a *method* for altering the method of representation. The history of the development of Dirac's theory of the electron is an instructive one in this respect also. In this example the analogies with Maxwell's equations and with the equation of relativistic motion of a particle in ordinary dynamics have been well exhibited by Frenkel. From the philosopher's point of view, an analogy helps one to surmount the ever-recurring difficulty which one experiences in getting accustomed to any new method of representation. Once one has seen in what respects the new resembles that with which one is already familiar, one is prepared to accept on trial its novelties and let the success of the application of the new method justify it. In course of time when a method has been tried and accepted, the analogy tends to be forgotten, and it is not necessary unless one wishes to exhibit the relation of one method of representation to the other.

The important analogies of physics deal with the substitution of one calculus for another. Being mathematically expressed the substitution is usually easy to comprehend. If one were to use mathematics exclusively in this connection, there would be few possibilities of confusion, but one has also to employ words to express thoughts in the usual literary forms which can be spoken, for, generally speaking, mathematical equations are not suited for speech. And for the following reason. In speech the mathematical equation has really to be described in words, and just as a complicated picture is not presented all together by a description, whereas a drawn picture is so presented, so the listener has difficulty in 'seeing' the mathematical form of the equation read out to him. This inescapable resort to ordinary linguistic signs, as opposed to mathematical signs, is one of the main sources of our philosophical problems in physics. We have to use words which play more than one rôle in language, and whose meanings therefore depend on their application; more confusing still there are some words which have not a sharp meaning in ordinary language, and some which have no regular function at all, such, for instance, as the word 'meaning' itself; in applying them to science we try to endow them with a regular function. This process commonly occurs in philosophical writing about science, especially whenever a change in the method of representation is being 'explained' in non-mathematical terms. The transition from the one method to the other leads us in talking of the second to take over words and forms of expression which have a clear application only with respect to the first. For instance, the 'st te' of a dynamical system, taken to mean a set of values of the momenta and of the coordinates corresponding to a particular time, is a clear conception which does not do violence to the logic of the use of the word 'state' in ordinary language. A system cannot be in two different states at the same time. But with the advent of quantum mechanics, the word 'state' was given a new meaning and it became necessary to explain the mathematical representation of one wave function as a linear combination of others by means of the so-called 'principle of superposition', as if to say that an atomic system can be in different 'states' at the same time (meaning the states of ordinary dynamics). There is an argument here in favour of changing our notation and of inventing a new term in place of this confusing word 'state'. So long as we restrict our thought to the mathematical calculation, everything appears clear and straightforward, but whenever we attempt to make physical analogies or models of the mathematical process, we are liable (but not *compelled*) to get into difficulties; and whenever one piece of the model is not joined to the rest we attempt to *force* them to fit into each other instead of carefully rearranging the parts of our analogy so that they fit properly as do the parts of a correctly completed jig-saw puzzle.

Perhaps the most striking instance of this process in modern physics is the bringing together of the particle and wave ideas with regard to the electron. Here is an excellent example of the piecemeal nature of the development of physical theories. On the one hand there is the dynamics of a particle and on the other the theory of wave optics. In the former we have as the essence of the idea of a particle, something at some place at some particular time. In the latter it is characteristic of the wave motion that we do not localise it like a particle; the essential wave properties depend on its being extended through space over a period of time long compared with the period of vibration. Of course a wave packet or group of waves has the possibility of representing the localised propagation of energy with the group velocity, and this supplies its logical connection with the particle method of representation, namely, that in a picture of coarse grain the two methods of representation give the same picture. Whenever we look on the matter in this way it appears quite clear and without logical difficulties, but if we think of the moving electron itself there is a difficulty. With one method of representation we may say 'the electron is a particle', and with the other method we are tempted to use the same form of expression and say 'the electron is a wave'. This is a source of logical difficulty in modern physics. These two statements about the electron do not go together because one of them postulates properties of the electron which appear to contradict the properties asserted by the other. Now one way of appearing to get rid of a contradiction is to legislate that it be disguised as not a contradiction and to invent a 'principle of complementarity' which, while it seems to allow us either expression as we see fit, is concerned really with permission to choose the method of representation which is appropriate to our needs.

Is it not rather striking that in ordinary dynamics we find no need for a principle of complementarity to enable us to represent the earth on the one hand as a particle without extension and on the other as a body extended throughout a sphere of 8000 miles diameter, or that in the dynamics of a gas we find no need of a principle of complementarity to enable us at one time to represent the gas as a continuous medium for acoustical problems and on another occasion to represent it as an assembly of molecules? It ought to be clear that in the older parts of physics we do not have any difficulty when we wish to change over from one method of representation to another. Physicists are accustomed to this process as an everyday affair. Yet in the description of the motion of electrons and other elementary particles, the change in the method of representation appears to cause a logical difficulty. It cannot be merely that the two methods have different possibilities of representation—we are ready to expect a difference between the assembly of molecules and the continuous medium. The difficulty in the present instance seems to arise from the form of words 'the electron is a wave', the electron still being thought of as a particle, and correctly so because the substantive 'electron' makes us treat it grammatically as we do other *things* in physics. Compare the sentence 'the electron is a wave' with 'the electron is an angular velocity' or with 'the electron is an aperiodic motion'. Both of these statements are clearly bad grammar. It is formally correct to endow a rigid body with angular velocity or a galvanometer coil with aperiodic motion, but we do not say 'the rigid body is its angular velocity' any more than we say that the galvanometer coil is its aperiodic motion. Now a wave motion is a process, not a thing, and an electron regarded as an object is not a process but its translation is. So if we must use 'is', we might try 'the translation of an electron is a wave' but clearly a translation is not a wave. It would be correct, however, to say that the translation of an electron is the translation of a group of waves without implying that an electron is a group of waves. What we mean by such a statement in the cramping form of speech into using which we have been trapped, is that the laws for calculating the translation of an electron are the laws for calculating the translation of a group of waves in a particular system. The present analysis does not dispose of all the difficulties associated with the wave method of representing the motion of an electron, and, it may be said, arising from lack of a sense of conscious management of the symbolism of physics.

Whenever one is clearly aware that one is representing and is using one method of representation *as opposed to another one*, one is likely to be careful not to make the mistake of mixing up the grammar of the two systems, or if one has been trapped by misuse of grammar into difficulties which are definitely logical, one is able to analyse the situation objectively and disentangle the proper grammatical usages of the one method from those properly belonging to the other. In the growth of a subject, however, when a new method of representation is in course of being developed, confusion of the old with the new is almost inevitable. Some men find one way out of the difficulties caused by the confusion, other men find other ways. There are those who dispose of the difficulty by a literary *tour de force*, and in the process probably create for less brilliant but more critical minds new problems where none existed before. Others are content whenever they have learned to manage the methods of physical calculation in the uncritical spirit found among students of elementary physics who render explanations without understanding. Familiarity with a difficulty will in the course of time lead one into the habit of ignoring it or, in order to dispose of it, tempt one to formulate a 'policy' for bridging the gap between the two opposed points of view. None of these devices will appeal to the man who wishes an elucidation of the grammatical situation which gave rise to the difficulty, and who having once become clear about the matter will strive to have accepted by other scientists a notation appropriate to the new ideas and not confused with the old.

In the foregoing pages it has been explained that in physics we choose the method of representation appropriate to our needs; that does not mean, however, that we are content to treat all methods of representation as on the same level of importance. A method which gives correctly the possibility of representing nature with greater detail than another, is thought of as the better method—and it is so when it is a question of showing detail. We think of the second method as an approximation to the first. We tend, therefore, to arrange methods in a series according to the degree of fineness of the pictures of phenomena that can be made by means of them, and of course from this point of view the last member of the series is the best. If this one has been in successful use for some years it is in danger of being looked on as the best possible. Indeed, all reference to the series of methods, if it ever was in the thought of physicists who think in the following manner, is forgotten, and the laws of the method of representation in question are thought of in the same way that the laws of nature have been regarded by mankind for hundreds of years.

Roughly speaking, the expression 'laws of nature' is based on the analogy with the laws imposed by a political sovereign on his subjects, except that when in physics we find what ought to correspond to the disobedience of a subject, we invent the alibi that we really did not know the correct laws of nature or we regard the phenomenon as an exception governed by a special rule, as if the sovereign had granted an indulgence. According to this way of looking on experience, so long as we have no experimental facts which break the law we are sure that we know the law of nature and are tempted to regard an infringement of the law as impossible. There is no need to labour this point, for every physicist knows of instances where the confident theorist has proved wrong in his predictions; at the same time one would perhaps have wished some of our leading experimenters had more frequently pointed out the error whenever a theorist has asserted that a particular experimental result *must* be wrong *on theoretical grounds only*. Only experimental physicists are competent to judge whether the result is experimentally correct *after* they have investigated the matter in the laboratory. As soon as well-authenticated evidence is produced, the failure of a law of nature causes us to change the law. Something that was hidden has been discovered, and very soon the new law takes on the authority of the old. Now if a law of nature can be altered what is the source of its authority? We have been misled by our analogy if we think that the correct law which we are groping to find by experimental investigation is *the* law imposed on the world and that phenomena *must* be governed by it, for what is meant by 'the correct law'? How are we to be sure that one hundred years from now, the laws of nature which we know and use may not have been given up and others substituted in their place? It seems that the expression '*the* correct law of nature' is not a proper grammatical expression because, not knowing how to establish the truth of a statement employing this form of speech, we have not given it a meaning. The analogy we see breaks down here, and the authority which we have been accustomed to attribute to a law of nature must be found in a place quite different from that suggested by analogy with political laws. Having been impressed by the fact that 'the laws of nature' have been subject to change, some writers have gone to the other extreme and proposed to dismiss, for example, the laws of conservation of energy and momentum as 'mere conventions', and apparently overlook altogether the difficult processes by which physical scientists came to these ideas.

It should be clear that the laws of mechanics are the laws of our method of representing mechanical phenomena, and that since we actually choose a method of representation when we describe the world, it cannot be that the laws of our method say anything about the world. If they do, experimental scientists have been misguided for three hundred years. What experience teaches us is that one method of representation is more appropriate than another in the sense that a map of the earth is more appropriate on the surface of a sphere than on a plane. The authority which we formerly attributed to the laws of nature in one way has now to be attributed in another to the logic of our method of representation, namely, in this way, that if we wish to make pictures of the world according to a particular scheme, then we *must* follow the rules of that scheme. This is not to say that the scheme determines what must be the form of the actual pictures we draw, but it does decide what pictures are possible. If distances on the map were literally interpreted as proportional to distances on the earth's surface, the plane representation of the earth's surface by means of Mercator's projection would allow some queer processes to take place on the earth, whereas a spherical map read in the same way would not allow those possibilities.

Thus what we have called the laws of nature are the laws of our methods of representing it. The laws themselves do not show anything about the world, but it does show something about the world that we have found by experience how true pictures of the world of a certain degree of fineness or of a certain simplicity can be made by means of the methods which we have learned to use. We *can* describe the process seen on a cinema screen, as if it were a continuous one, provided that we make the rule that observations are not to be so refined that we can show the actual succession of images of discrete pictures. We *can* describe the motion of the particles of atomic physics by classical dynamics provided that the dynamically significant fields of force in which they move are not given structure on too small a spatial and temporal scale. We *can* describe the motion of the planets round the sun over a few years without introducing relativity and so on. In every one of these cases, the structure of the facts being represented is shown in that the particular method of representation adopted in order to describe them is successful or appropriate *in a particular way*.

Our choice of a method of representation is a real choice; we are guided by our experience in finding it. Once we have adopted a method and new facts appear which do not fit at once into the scheme, we again have a choice—either to give up the former method and use a new one which does have a place for the new facts, or else to adopt a special hypothesis to enable us to retain the old method. Either alternative may have the consequence that certain possibilities are created, whose correspondence with reality can be investigated experimentally. If the facts are found to disagree with these possibilities, we are free again to invent another hypothesis or to start once more with a new method. Generally speaking, we proceed in science like the mechanic who is continually adding gadgets to deal with this or the other defect in his machine. Sooner or later the gadgets weaken the whole structure, and he is well advised to make a new engine the design of which incorporates as essential to its structure the law of functioning which removes the need for the gadgets. In science the process that corresponds to the designing of a new machine is the invention of a new method of representation. Once it is adopted the structure of the science in which it is to be used appears to have been tidied up and loses its patchwork appearance. A new method may have wide application, in which case its invention will probably be attended by philosophical discussion concerning its 'fundamental' importance for human activities which certainly do not enter into the profession of physics; on the other hand, the new method may appear to clear up only a small part of the subject, and if the change is noticed at all by others than specialists closely interested in the matter, it is accepted as the customary thing. The development of any new experimental technique is of this character; experimental facts are simplified and made clear when the proper procedure is followed, as for instance in the science of the photoelectric emission from metals and of the photoconductance of insulating crystals.

It should be made clear that a hypothesis added to a method of representation is not *necessarily* an excrescence—only experiment can decide this. The discovery of the neutron is an excellent example of a successful hypothesis which easily finds a place in our theories; it was based in the first place on what at the time seemed the curious failure of other possible mechanisms to explain the properties of the radiation observed in certain atomic transformations. In contrast, the neutrino which has been invented to account for the lack of energy balance in β-ray changes has the appearance of a much less appropriate hypothesis because up to the present the properties postulated of this particle seem expressly designed to exclude the possibility of its detection in ways analogous to those used for the other particles of atomic physics. The question, however, is not to be settled on the grounds we have just mentioned. It is for the experimental scientist to discover whether or not there lies before the hypothesis a useful life or a rapid disappearance into oblivion. We have to learn physics a little at a time, and there is no good purpose served by refusing to give a hypothesis a fair trial merely because one feels that it does not fit easily into our present scheme of things; one may be right as judged from the point of view of the distant future, but wrong in one's judgment as to the way in which the goal is to be reached. Think of the introduction of the hypothesis of quanta into physics. It has brought twenty years of discomfort in physical theory and ended by developing into a full-grown method of representing atomic phenomena which derives from the systems of mathematical knowledge that have supplied its calculus an impressive appearance of logical completeness just like classical mechanics. This is not to say that all the problems of atomic physics have been solved, but to emphasise that we now have a method of representation which has the scope of certain well-defined branches of mathematics; its alliance with them gives to the physical theory the appearance of being well established in its 'fundamentals'. Compare the present mathematical form of quantum mechanics in the hands of Heisenberg, Dirac, Schrödinger and Weyl, to mention only a few outstanding names, with the form given to classical mechanics by Lagrange, Hamilton and Jacobi. The Hamiltonian method in ordinary dynamics involves the assimilation in the subject of the whole theory of systems of linear differential equations of the first order and their treatment by means of the solution of a partial differential equation. Since we have the possibility of choosing our methods of representation it is good to take advantage whenever we can of the symbolic systems with which mathematics provides us and to use them to give a logical backbone, and what is just as important, a good notation to our physical theories.

Having done this we have to remind the mathematician where the appearance of infallibility originates in his method. It is because we have chosen his method and are prepared to support it with special hypotheses to cover any facts which are not at first correctly described by the method; we are little disposed to give up using the method until, as the scope of experimental investigation widens, the patches become so numerous and disjoined that one begins to look for a new method of representation. Then the process of growth usually repeats itself, the mathematical theorist falls in love with the new method, and the experimenter gradually accumulates the knowledge which may lead sooner or later to a change in the method of representation or to the invention of a new one where none existed before.

We build the theory just as we develop experimental technique. At first there is the trial with crude temporary devices in our apparatus; how crude they may be every experimenter knows. Once such a device is made to work we incorporate it in our apparatus and technique—we trust it and depend on it. It is so with hypotheses. Tentative at first, they become the foundation of new developments in our theory; we trust them as long as they will work, or accept them for what they do and make up for their deficiencies by additional ones which in turn have to go through the same process. We can never tell when we shall have to give up a theory any more than we can tell before we have used it when an experimental technique has to be abandoned. A theory or a technique becomes infallible only when we agree to make it so by a convention.

The preceding discussion shows us the so-called Principle of the Uniformity of Nature in an interesting light. Studies of the pleochroic haloes in mica having supported the hypothesis that they were produced by the various groups of α-particles emitted during the radioactive transformation of uranium and the products of its disintegration, some men have wanted to say that if the radioactive theory of their origin is correct, these haloes, which must have existed for some hundreds of millions of years, provide evidence that the laws of radioactive transformation have not changed during that period of time. In order to determine whether such a statement is a proper statement about the world, all we have to ask is the question, what if the haloes had a different structure? Would we then still hold to the theory that their origin was a radioactive one? Clearly we could not without the aid of an additional hypothesis to cover the disagreement between the structure of the haloes actually observed and that structure predicted by the unaided theory of their radioactive origin. Suppose that the hypothesis advanced is that the laws of radioactive transformation were different in the past; one would then look for a physical cause of the changed behaviour of radioactive substances. How could one be certain that the cause operated in the distant past according to the laws established by experiments made in the present era, and so on? Now this is just the kind of philosophical difficulty which was disposed of by geologists last century and which has never really worried astronomers. The Uniformity of Nature is not a hypothesis about the world at all—we are not prepared to substitute an alternative hypothesis—it is a statement concerning our method of representing nature. We choose to represent nature in this particular way, and for the very good reason that, if we do not ascribe to matter the same properties that we can measure now, we have nothing to guide us in the choice of a method of representing the changes that are supposed to have taken place. Suppose that it were seriously contended that a hundred million years ago the solubility of common salt in water was twice as great as we know it to be to-day and expect it to be to-morrow. Every physical scientist would expect to have an explanation of the changed solubility in terms of the kind of theory to which he is accustomed. If this explanation is not forthcoming he is certain to dismiss the idea of the doubled solubility of salt in the distant past as the kind of notion pleasant enough in literary invention, which is useless, however, from the scientific point of view, because the change proposed in the symbolism of science cannot be employed in calculation. The moment it is asserted that the past solubility was twice as large as at present, one can at least conceive the possibility of some time when it may have been 1·5 times as great as now, and so on. That is, if we employ a numerical ratio, functioning as a proper arithmetical symbol, it is logically correct to regard it as a variable whose particular value is determined by physical causes. Otherwise we say that the number in question is a constant of nature; in which event, of course, the logical possibility of substituting one number for another is retained until we have agreed on a convention as to which value is to be used, after a number of careful measurements have been made. Indeed, we need to retain the possibility of altering the value given to the 'constant' in the light of future experimental knowledge. Examples of this are to be found in modern physics from the charge of the electron to the number of wavelengths of red cadmium light in the international metre. But these numbers are not represented by variables in the theory.

Whenever one sees any principle of a general type such as the Uniformity of Nature put forward in science, one should recognise that it is a principle regarding the method of representing nature that is involved. The appearance that one knows any principle of nature and can decide on its correctness without any reference to scientific experiment (as opposed to commonly held vulgar notions regarding experience) is given by the form of expression 'So-and-so Principle of Nature'. Such principles give some philosophers an easy target when they think they are attacking the logical foundations of science, while other philosophers make a business of inventing them. As a matter of logic, what upsets the philosopher is merely the notation but he does not know it, whereas the scientist, whenever he wishes to do so, changes the method of representation. In physics there have recently been invoked by theoretical writers two principles which are clearly applicable only to the method of representation. First, Heisenberg has adopted 'the Principle of Symmetry in Nature', and secondly, Born has appealed to 'the Principle of Finiteness in Nature'. It is certainly not without importance that two men of such outstanding achievement should appear not to know what they are doing when they appeal to such principles. Notwithstanding the possibility of referring to more than one principle of symmetry or to alternative principles of finiteness, it is perfectly legitimate for these writers to have adopted the particular principles which served their theoretical purposes. But it must be admitted that the form of expression 'Principle of Nature' at first glance has an appearance of finality—as if one should say 'this must be conceded: everyone recognises this', when as a matter of fact it was the genius of Heisenberg and Born that recognised the fruitfulness of the particular form they had chosen as an integral part of their respective theories. The position of such principles in theoretical physics is quite comparable with the position of axioms in geometry, except that mathematicians have known for a long time that they are free to choose systems of axioms, whereas theoretical physicists as a body do not seem willing to admit that they have chosen to use principles such as we have mentioned (axioms they might be called) as the expressions of essential logical form in the theory they have in mind; instead they try to make it appear that there is no choice about the matter—and what better can one do than blame nature! We have here touched on a confusion of thought which is somewhat complicated. Why should it seem necessary to disguise the introduction of a new idea in theoretical physics in a form of presentation which appears to endow it with unquestionable correctness? Is it because of a convention as to how new ideas are to be presented? Is it an attempt, so to speak, to coat the bitter pill? Why should theoretical writing be expected to provide its own justification, when indeed the justification of the choice of a particular theory (as opposed to other possible ones) lies in comparison of its predictions with the relevant experimental facts? And finally, why is it that once a theory has been expressed in the form conventionally adopted by mathematicians in expounding their subject, some men cease to have any doubt as to its correctness as a means of predicting physical phenomena? We shall not attempt to give any answers.

In astronomy the laws of dynamics are employed to calculate the motion of the solar system. To use in this connection the expression 'change of the laws of dynamics', meaning that in the past the laws did not have the form we now know, is a mistake because, if we have in mind a different law, we must know what the law is in order to calculate, or we must know the law of change of form of the laws of dynamics. In either case the net result would be that the calculation would be performed in terms of a new system of dynamics (we are not attempting to consider relativity in this connection) which would require that the laws of motion are not invariant under the transformation $t' = t + a$, which changes the origin of time. Such a law can be treated as if at any particular era the laws of motion are invariant with respect to this transformation, provided that motion only during a comparatively short period of time is in question. This situation would be analogous to the description of the orbit of a planet round the sun as an ellipse whose determining parameters change by a small amount on each complete revolution, when in fact the planet describes a continuous curve which is not a succession of ellipses, but which can be fairly well described as such. It is rather striking that the difficulty we have been considering has arisen with regard to time and does not seem to have occurred as a serious question with regard to separation in space. No one has proposed that the laws of dynamics are different at different parts of space; indeed one does not need such a hypothesis; dynamics leaves a place for connecting motions at different points of space by means of the potential energy function.

We have referred on several occasions to the use of mathematics in physical representation; we now return to it. Lenard in his recent *Great Men of Science* has made some interesting remarks on the place of mathematics in scientific research. Anyone familiar with the opinions of experimenters will recognise in the following words quoted from this book* a common evaluation of mathematics in relation to experimental science.

...mathematics is, throughout scientific research, simply a tool, enabling us to apply knowledge derived from the observation of accessible and generally simple cases, in a completely logical manner and without danger of mental error, to all other cases whatsoever, no matter how complicated; but knowledge of nature can only be derived from observation.

It is the special virtue of the art of mathematics that, when correctly applied without arbitrariness, it keeps away all foreign matter and only allows that to take effect which was originally derived from observation and put into the equations....

...Mathematical skill can, of course, bring about advance in scientific knowledge; but this only happens when the discovery and observation of natural processes which are still unknown, or of a new kind, or not yet properly understood—the highest achievement of the experimental investigator—yet fail to bring full understanding of the processes by means of simple considerations, by reason of their complexity.

Thus Newton became by means of mathematical skill the discoverer and founder of the law of gravitation, by examining the complicated motions of the planets, and calculating these down to their details. But the description we have given, which was

* Pp. 220 *et seq.*

entirely limited to essential points, must have shown the reader in this case also that the discovery itself—though not its confirmation in all directions—nevertheless resulted from comparatively simple lines of thought, which also needed only fundamentally simple mathematical means for their support....

Here Lenard tries to express something concerning the logical connection between mathematics and experiment; it is of some importance, although the above statement is hardly likely to influence anyone who is well trained in mathematics and who understands the history of physics. What is meant by 'simple considerations'? Is there any clear division of mathematics into simple and not simple, that is not relative to some standard of judgment of simplicity? Would one, for instance, call the use of logarithms simple, having regard to the labour of the calculation by which the tables of logarithms were constructed? In some geometrical problems, the calculations by Cartesian analysis are expressed by complicated symbols, whereas the corresponding calculation by vector methods is expressed in simple form. Is the simplicity of the vector calculus in such a case to be the deciding factor, or do we say that the geometrical relationship in question is a complicated one because its Cartesian form of expression is not simple? Quite clearly the degree of complexity of a calculation depends on the appropriateness of the symbolism used. By inventing symbolism, mathematicians often provide us with methods for the formally simple expression of complicated ideas, as in tensor calculus. If a physicist is really familiar with the appropriate mathematical method, he will regard the statement $a=bc$, where a, b and c are ordinary numbers, as no more simple than $A=BC$, where A, B and C are matrices. Lenard, although he almost appears to say so, cannot have meant as a standard of simplicity that one is to be allowed, for example, arithmetic, algebra, trigonometry and the first six books of Euclid, but not the differential and integral calculus, for any such distinction is so arbitrary as to be pointless. What ought to have been expressed by Lenard is that theories must in the first place do for experimental science what a crude map does for a surveyor: it acts as a guide for more exact work. If the map does not show the main landmarks it is of no use, and likewise a theory which does not show by even the most elementary mathematics the main facts which are obtained by means of experiments of a relatively crude type has no claim for serious consideration as the method for calculating effects open to study only by more refined methods of observation.

Let us now examine the contention that mathematics allows to take effect only what was originally derived from observation and put into the equations, and consequently does not add to our knowledge. Mathematics does not perform experiments for us to inform us of facts, but because of calculation we make it possible to test the theory in a more searching manner than without it, and indeed in the investigations suggested by theoretical prediction we may be led to discover places where the theory appears to fail. Lenard seems to suggest that all the consequences of the mathematical theory must agree with the facts because the theory merely takes over relations established in the laboratory. This need not be so.

There is no clear division between the mathematical theory so-called and what the experimenter does. The mathematical theory is not a representation of the experimental laws, it is the expression of them; the experimenter employs the same mathematical connections in stating the laws he has found by induction. Consider the emission of electrons by a hot body. The dependence of the current on the temperature is given by a formula which is not simple, and the choice of the particular law used in preference to other empirical laws which might be used to fit the data of experiment is governed entirely by theory. The experimenter is guided by theory in his choice of the mathematical relationship between current and temperature, and this leaves on one side another important matter, namely, that the current depends on the electric field in the neighbourhood of the emitter, and here his treatment is again guided by theory. In such a case as this the mathematical theory is not something superposed on the experimental facts, as if we knew the laws from experiment then expressed them mathematically; as a matter of fact mathematics got into the laboratory in the nude state. To distinguish alleged non-mathematical experimental results from the so-called mathematical theory is entirely artificial and shows ignorance of how language functions. The results of Faraday's investigations of electrostatics and electromagnetism were presented in non-mathematical terminology, and were expressed by Maxwell in explicit mathematical form. Is one to say then that Maxwell created the mathematical form which represents Faraday's results? No. Maxwell translated Faraday's thoughts into different words, namely, the symbols of mathematics; the logical form of thoughts is unaltered by such translation if correct. Either the logical form of the mathematical relationships used by Maxwell was present in Faraday's thought of the phenomena or else Maxwell invented the form, in which case we should have no longer any right to speak of a mere translation of Faraday's experimental facts. Maxwell did of course make his own contributions to electrical theory. The contrast we have been considering is comparable with that between the description of a situation in space by words and its specification by mathematical equations.

One cannot expel mathematics from the laboratory, but one can refuse to deal with complicated mathematical calculations which are necessary only to present details that are beyond the reach of our most refined methods of measurement. In this connection, consider the equation of state of a gas. As the first general statement of the law we have $pv=RT$, and this is satisfactory until sufficiently precise measurements over a wide range of temperature and pressure lead us to Van der Waals' equation, for example. The application of quantum statistics has yielded a form of the equation of state with still greater detail of representation. Are we justified in asserting that this newest law is *the* correct one rather than either of the others? The only test of correctness is comparison with experiment, so the word 'correct' as applied to a physical theory has to be understood as 'correct relative to a certain degree of fineness of the observations'.

Once we learn to look on the matter in this way we shall put behind us the temptation to endow any theory with absolute correctness and shall be less willing to criticise as incorrect from the theoretical point of view a formula such as $pv = RT$, without having in mind the manner of applying the equation to experimental facts. This may seem a small thing, but insistence on it will probably make less generally held the attitude to physical theories which enthrones the most elaborate theory we know as the essentially correct one.

The experimenter cannot escape the logic of his methods of representation. Suppose that in the laboratory one is required to connect in the form of a certain electrical circuit, for instance a Wheatstone bridge, certain instruments, resistance boxes, etc. If one is not accustomed to do this, one will draw a diagram of the circuit, name the connected points, write the names of instruments or other pieces of apparatus opposite the corresponding arms of the bridge in the diagram, and then follow the directions for connecting the apparatus just as one follows any other instructions presented by a diagram and names. However, this is not necessarily how one proceeds who is familiar with the process. He does not need to draw a diagram. He may imagine a diagram, but even that is not necessary. He has enough signs in the actual things and wire to be joined together to make the 'diagram' with the connected network. To an experienced worker it is just as easy and direct a process to connect the network and to see that it is correctly wired, as it is to make a diagram and see that it is correctly drawn. Imagine that someone arrives in the laboratory when all the connections have been made, he recognises it as a Wheatstone bridge merely by looking at the connections. If a mistake is made in recognising the circuit it is not even then necessary to use a diagram of the circuit, for the difference between the actual network and the other can be shown by pointing out on the bench how the connections would have been made differently if the other bridge had been intended. The logical situation is the same as that in music when one sings a song without having before the mind's eye a picture of the whole tune, such as a visual image of the score from which it was learned. Mathematics gets into physics in a manner not unlike that in which the musical form of a tune gets into the process of singing it.

Of quite a different nature is the application to physics of mathematical theories in a manner which suggests numerology. The essence of these theories is that a more or less elaborate mathematical scheme is proposed, and whether or not it is dressed up in a physical model, all that the theory predicts that can be compared with experiment are numerical constants—usually the non-dimensional constants of nature. The point of view from which such theories are advanced and supported is that all that experimental science does test in our theories are numbers. But here a mistake is made in that one has overlooked that the numbers of ordinary physical theory are variable numbers. It is true that when one gives a particular value to the volume of a definite mass of gas, the pressure can be calculated if the temperature is known. When v and T and m are fixed, then p has a definite value and the prediction can be tested by measurement. But the importance of this calculation in exemplifying the method of physics is that in it one applies the law by which the pressure can be calculated under any other conditions we care to specify. The law connects variable numbers.

When one is dealing with a theory which purports to show why a constant of nature has its measured value, one has a right to ask what is represented in the theory and on what grounds the chosen mathematical structure is relevant to physics, for clearly we can imagine a particular number being produced as the result of an endless variety of mathematical processes. Relevance is established logically only by the appearance in the theory of laws that can be compared with fact. Any other connection such as that by the use of physical terminology in speaking of the mathematical structure is not a substitute for this logical connection by identity of structure in some respects between theory and fact. If it is asserted that the theory can be compared with experiment in no other way than by means of the values to be assigned to the constants of nature, the theory is useless for experimental science, and its logical value is likely also to be negligible.

A constant of nature is a fixed number, not a variable number. If this number is to be fixed by a theory, the function of the theory is to show how in a calculation by means of it a variable number is not to be given on one occasion one value and on a different occasion another, but is to have the one particular value found by measurement. The theory shows the constant as a particular value of a variable, but the possibilities from which it makes a choice do not represent actual physical possibilities of nature, for the theory pretends to establish that they are not.

It is always possible to think of an explanation of a physical law by means of a more elaborate theory which picks out the law in question from other possible laws. For example, the kinetic theory of gases explains Boyle's law. But this process must come to an end just as soon as we are unwilling to consider as relevant to the description of nature the system of possibilities from which the theory has to make a selection. The only test of this relevance is that the representation we have in mind is that required to show experimental facts connected by a coherent theory. The molecular theory of gases was at one time objected to on this very ground, but as soon as Dunoyer demonstrated the properties of a molecular ray, that attitude had to pass away. It is pertinent therefore to ask of any theorist who proposes a superstructure to physics in the form of an explanation of what we accept as fixed things of our methods of describing nature, that he show some place where his theory can be compared with experiment in addition to those respects in which he 'explains' what already is accepted.

CHAPTER IV

THE NATURE OF MECHANISM

IN its dictionary sense 'mechanism' is the structure of a machine, and in this sense it will be used here; but in case the reader has restricted ideas about the meaning of the word 'machine' it is well, perhaps, to mention that we talk of 'the mechanism of muscle action', 'the mechanism of the ear', 'the mechanism of the nervous system'—to give examples from physiology. You understand also what is meant when I refer to the mechanism of the flight of a bird, even although you may be ignorant of aerodynamics or of the anatomy of birds. On the other hand, when one speaks of the mechanism of chemical action, it may not seem clear how this use of the word accords with the dictionary meaning; but since the word is used in this way, we have every right to expect that the mechanism of chemical action has something in common with the mechanism of a lathe, for example. This is the way in which the problems now selected for discussion are to be approached. What there is to be said here of importance is so because it does not involve detailed technical knowledge of particular machines or particular processes. We are concerned with the logical essence of mechanism.

If one were asked to explain how a watch works, one would open the case and point out how the oscillating wheel controls the release of energy from the mainspring through an escapement and a series of geared wheels, and one would make statements such as: 'This wheel engages with that and turns three times as fast, for it has only one-third of the number of cogs that the other one has.' By the mechanism of the watch we mean the arrangement of wheels, levers, and so on, functioning in this particular way. In our explanation we point to the connection of each part of the machine with other parts. When we understand these connections we see that the parts *must* move in the definite manner which is characteristic of the particular machine. Whenever we think of an actual machine, it appears *necessary* that when one part of the machine moves in a certain way, other parts should move in their proper manner. On the other hand, we can form a clear idea of a machine behaving otherwise than according to the law which at another time appeared necessary—for instance, the engaging cogs in two of the wheels of the watch might slip. It is therefore not impossible for such a thing to happen. We have then to distinguish the theoretical geometrical picture of the machine from that obtained by actual inspection and measurement of its parts. Once we recognise that the cogs are worn to the extent kinematically required to permit the slipping to occur, we admit that the watch is no longer compelled to function in the way which we had previously regarded as inevitable. Wherein, then, lies the necessity that the machine should behave in the way it does? What is the source of the necessary connection between the motion of one object and the motion of another?

To the ordinary man it appears evident that two bodies which touch each other can act on each other affecting each other's motion. Nothing is more obvious than the deflection of two balls when they strike one another. It is also evident that when one thing is joined or connected to another by means of matter of any kind, these things can affect each other's motion reciprocally. But when it is suggested that action may take place at a distance, it seems impossible to some men that such action is anything but a mathematical fiction, and that the two places are not connected by some hidden matter. They point out that so long as we suppose that action is a property of the thing itself, we are faced with the puzzle as to how it acts where it is not. Accordingly they invent objects which are everywhere—the most famous example of which is the aether by which electromagnetic effects are to be propagated from place to place—and the invention is supported not on the ground of its scientific usefulness as a hypothesis, but on the logical necessity for its existence, in the following words of Sir Oliver Lodge:*

* *Modern Views of Electricity*, p. 362.

Now I wish to appeal to this mass of experience, and to ask, Is not the direct action of one body on another across empty space, with no means of communication whatever—is not this absolutely unthinkable? We must not answer the question offhand, but must give it due consideration; and we shall find, I think, that wherever one body acts on another by obvious contact, we are satisfied and have a feeling that the phenomenon is simple and intelligible; but that whenever one body apparently acts on another at a distance, we are irresistibly impelled to look for the connecting medium.

Even the continuous medium will not do for us what Lodge—and others, of course—think it will do. The medium's function is to connect one part of space with another, and this appears to be assured because the medium is continuous and any part of the medium touches parts adjacent to it. Clearly, however, adjacent parts are not in the same place, so there still remains the problem of connecting them; but this is usually overlooked, as is also the thought that the differential coefficients, which express the state of the medium at a point, properly belong to the point *and* its neighbourhood. Thus the connection of two places close together is essentially the same matter as the connection of two places far apart.

Is there not something in common between connecting things and parts of space and the connection between the musical score from which the pianist reads and the notes he touches on the keyboard? Indeed it is a commonplace to speak of playing the piano as mechanical. Machines have been made to play the notes according to the score inserted in the machine. So also do we speak of the mechanical character of arithmetical calculation, and of mechanical dexterity in dealing with algebraic symbols. There is evidently something in common between calculation, reading and the necessary connections which we see between the parts of a machine.

Two hundred years ago, Hume first gave a clear insight into the nature of necessary connection. He pointed out that we wrongly ascribe the necessary connections to the things themselves, whereas "When we say that one object is connected with another, we mean only that they have acquired that connection in our thought." In spite of this clear statement, in physics, up till less than a century ago, so much was the necessary connection associated with ordinary matter, that there was a well-established willingness to explain every phenomenon in terms of wheels, levers and so on, or at least in terms of rigid bodies which it was postulated we cannot see. The man whose work put an end to this was Clerk Maxwell. In his theory, electric and magnetic fields at different places and times are connected by certain laws without any reference to the old-fashioned systems of connection by wheels, elastic solid and the like. His theory initiated a revolution in the style of physical theory. For the first time in the history of physics, mechanism was consciously accepted in the form of mathematical equations. Of course, the revolution is not yet complete—after all, we are removed from his time by two generations only, and the pre-Maxwellian tradition is with us still by sheer inertia. Modern physical theories, however, are in the direct line from Maxwell, in that they boldly express connections by means of mathematical equations.

What is it about a mathematical equation that can present mechanism? The signs we make are not tied together by cords, nor do they collide with each other. The answer is, of course, that by mathematical equations we express laws of connection. For example, consider the simple machine composed of a weight A connected by a string over a fixed pulley to a trolley B mounted on horizontal rails. What is essential about the simple connection of A and B we express mathematically thus: $x+y=a-c$, where x and y denote the respective distances of A and B from the pulley, a is the length of the string, and c is the length of that portion of it which rests in contact with the circumference of the pulley wheel. That is, the possible values which we might give to x and y are limited in our thought of this machine by this equation. The equation expresses the law of connection for anyone who understands it. In this particular case it is a law of kinematic connection.

Let us consider now the dynamical aspect of this simple

machine. We say that the weight A causes B to move: B always moves when under the action of A. We cannot think of B not moving under the influence of the weight connected to it without manufacturing automatically the need for a more complicated description in which again there is a rigid causality (agency of causes). For instance, we have sooner or later to introduce the idea of friction. In experiments we find that force measured in terms of the hanging weight is not equal to the product of the mass of the moving parts and their common acceleration. In order to retain the law $F = ma$, we are compelled to invent frictional force and to discover from experience the laws of friction. It is important to notice what has been done in this process. In the first place, using Newton's laws and taking no account of friction, we predict that A's weight W causes B to move with the acceleration W/M, where M is the total mass of A and B. Secondly, we observe that the actual acceleration is different from this. Now, we argue, B must be caused to move in the way that it does move, and so we alter our description of the machine and introduce friction (and whatever else may be required), saying that the motion of the system is now under the action of two (or more) forces. The calculation by means of the dynamical law may now be in agreement with experiment. If it is so, we expect it to agree whenever we are dealing with the same or a similar system. If the prediction is not in agreement with facts, we have to look for a new cause of the difference between theory and experiment. Such, for instance, would be the existence of electricity and magnets and their mutual action in the system. But if we wish to use this as the explanation of the observed motion we must be able to describe the motion in the special way demanded by the theory of electrodynamics; for the hypothesis of electrodynamic action will be subject to other tests. It is not enough that this hypothesis has the possibility of the observed motion it was introduced to explain; it must be able to account for other possible motions, such, for instance, as those produced in electrical instruments used to measure the electric charges and currents supposed to cause the deviation from our former prediction. Suppose that the measurements with the electrical instruments do not support the hypothesis that the motion of the system we have been studying is influenced by electro-dynamic action, we then look for another physical cause, and, if necessary, say that the cause of the deviation of the original prediction from what is observed is a new physical phenomenon. The invention of the so-called blocking-layer photoelectric cell followed the discovery of such a process. Experiments with copper-oxide rectifiers revealed the action of an unexpected cause which was found to be the photoelectric effect of the light incident on the surface of the copper oxide. The same principles operated in the discovery of cosmic rays from studies of the natural leak of electric charge from insulated charged bodies.

There is no necessity that nature will be observed to follow our expectation of its behaviour according to our theories. We learn by experience how to describe what appear to us to be similar situations and processes. If our predictions are not fulfilled, we look for the differences which necessarily exist between the new and what we are familiar with, and ascribe these differences to the action of the appropriate physical causes. Nevertheless, one does feel that there is necessity somewhere in the functioning of every machine and in every physical process. The philosophical elucidation of this problem of necessity in relation to explaining the action of machines lies in this, that in our thought of the machine we transfer from the calculation to the machine the inevitability of the answer to the calculation of our expectation as to how it will behave. Thus what we call causality does not correspond to a compulsion on nature to go in a certain way, but corresponds to the correct use of our symbolism. We choose a definite method to represent what will happen, based on observed resemblance in some particulars only between the process to be studied and processes which have been studied in the past, or it may be with imagined processes whose law we know. Once the method has been chosen, we are committed to use it, and the answer to any properly chosen question will be given by correct application of the method, with the same certainty that applies to the correct use of arithmetic. The *necessary* result of the calculation by the method chosen is the answer to the particular question asked. The answer is correct, and what we should have in mind when we say this is its logical correctness with respect to the application of the chosen rules of calculation. Whether or not the answer is correct in this respect can be decided without experiment. What we cannot decide without experiment is whether the answer to our question represents not merely a possible way in which the machine may act, but the actual way in which the actual machine under observation does act. If the prediction is correct in this experimental sense, one is tempted to confuse the certainty of our knowledge that the prediction was obtained by correct use of our symbolic method with the agreement of prediction with observation. Our symbolism requires the picture of the machine to go in this particular way, the machine is correctly represented by the picture, so one is tempted to say that since the behaviour of the machine follows the law of the picture, therefore the machine is compelled to follow the law, as if the future performance of the machine were determined for all time. In an actual machine, a cogwheel may become worn or an electrical connection may become bad and so on, and one is inclined to dismiss these as accidents, unless one has made a study of the wear of machinery and the deterioration of electrical gear, in which event one does not think of them as accidents but as due to the action of natural causes. The fact is that only the lapse of time can show us how successfully we represent the machine—anyone who has constructed an apparatus for experimental research is familiar with this truth. Machines wear out, are broken by accidental and intentional human agency, and, generally speaking, in practical affairs one does not have in mind the application of the physical calculation regarding a machine at times later than the incidence of such accidents. The representation of the machine is a short-term one. In astronomy, however, when applying dynamical laws, we have had to postulate a great age for the system of which that subject treats, and what is just as important, one seems to see the possibility of a long-enduring process in the future, determined according to the same laws which we know to-day, or, at least, according to some laws which do not differ greatly in their short-term consequences from those we know. To the question 'Will the sun rise to-morrow?' we answer an unhesitating 'Yes', and this is the proper answer if we understand by the answer that it is the expression of our expectation. If, however, we understand by our answer that there is no other possibility but that the sun must rise to-morrow, we are misusing language. The very fact that language allows us the possibility of saying 'The sun will not rise to-morrow' asserts the possibility of conceiving that such a thing might happen. And if it did—assuming that the catastrophe left some physicists able to continue their scientific activities—these men would have to find a physical cause to explain the phenomenon, and they might even be willing to change drastically the very laws whose correctness one appeared quite unwilling to question when one answered 'Yes'.

The process of finding a physical cause to explain the disagreement between our prediction or expectation regarding a phenomenon and what actually happens, is well understood. One proposes hypotheses and examines them in turn. For instance, one sees what appears to be an electric-light bulb in a socket, but does not feel the warm glass when one extends the arm to touch it. One makes the hypothesis that he has seen the image of a lamp in a mirror, and tests the hypothesis by looking for the mirror and the actual lamp. On the other hand, the logical status of the so-called law of causality is not clearly understood. By this term is meant the general necessity that every effect has a cause.

We have already discussed how any physical phenomenon may either fit appropriately into the relevant scheme of physical representation, or it may require us to introduce special additional hypotheses to explain it. Only in the latter case is it correct to speak of the cause of the phenomenon in question. For example, the planets move round the sun and are caused to do so by the gravitational attraction of the sun, only relative to our system of representation by ordinary geometry and Newton's laws of motion. This system allows the possibility of other modes of motion. Whereas, if we substitute geometrical laws for the dynamical ones and choose for the representation a space-time in which the actual orbits are determined by the principle of geodesic paths, we cannot properly say that the sun causes the planets to move in the way they do, because in our representation there is no possibility of their having a different path which does not violate the geodesic law. We may, however, say that some agency causes a deviation from the geodesic path. It would be correct also to point out that the chosen method of representing the solar system shows in its application what in the Euclidean picture is spoken of as the gravitational action of the sun in determining planetary motions. Let us consider another quite different example. Suppose that one has to measure the period of a pendulum by means of a timing apparatus in which one makes measurements of length on a strip of paper. According to the criteria of experimental practice as to the possible error of measurement by means of the chronograph, one decides that the numbers which are given by calculation in a series of determinations of the period and which agree among themselves within the possible error, represent the same single possibility for the period of the pendulum according to the measuring device used. Relative to the system of representation of the degree of fineness chosen for the measurement, there is no physical cause for the different numbers obtained by calculation from the measurements of length and so on, for these numbers do not represent alternative possibilities by the method of representation employed. On the other hand, if a physically significant difference in the period is observed, it is quite correct to ask the cause, for we have the possibility of testing by means of the system of measurement provided, the effects of different causes such as change in temperature, change in the acceleration of gravity and so on; the possibility of testing is shown in our symbolism as the possibility of representing. Unfortunately we are very easily led away from this way of looking on experience because it is usually possible to improve the measuring apparatus so as to obtain more refined measurements. But so long as we keep in mind that the method of measurement is part of the symbolism in that it provides the means of testing the correctness of any picture of nature belonging to the system of representation adopted, we shall avoid this temptation. The situation is just that found in regard to the meaning of a word: it is the logical application of the word in language to represent some state of affairs that determines the meaning. Meaning is connected with the *method* of testing the truth of a statement which employs the expression whose meaning is in question. It is so in physics. The exact meaning to be given to the word 'length' in a statement such as 'The length of this iron rod is 25·30 cm.' depends on the degree of precision of the method which one has in mind for measuring it.

We have attempted to make clear how a cause operates relative always to a system of representing nature. When we say that every effect has a cause we think that we are attempting to say something about the world, when indeed what really concerns us is the logic of representing nature. The 'effect' is represented when we describe it, and in describing it we point it out as one possibility as opposed to others. What corresponds to the necessary connection between effect and cause is the logical necessity that in order to pick out one from a number of possibilities, one must be given the particular law for choosing it. This is just what the physical cause does—it provides us with such a rule. The change of the resistance of a piece of wire is supposed due on some occasion to the change in its temperature; this temperature change 'causes' the change of resistance, provided that the actual resistance agrees with that calculated according to the law of temperature variation of the resistance of the material of which the wire is made. If one wants to say that the law of causality does not hold, one is misled, for what could it mean except that, there being no cause, no rule exists for selecting the possibility which is the effect alleged not to have a cause? That is, the effect in question cannot be described; which is nonsense. The law of causality is not a law like the inverse square law, for we can say what it means not to follow the latter. It has perfect sense to put in opposition to the law of the inverse square of the distance that of the inverse cube, but what does not follow the so-called law of causality cannot be described, and we do better not to speak of a law at all.

Before proceeding to discuss the application of the analysis just made to the view frequently expressed that quantum mechanics has abolished causality in science, we must draw attention to some other important logical questions in this connection.

In thinking of the agency of a cause one usually imagines that an event yet to take place cannot be the physical cause of a process observed now. It is quite certain that no human action can change events that have happened (logical impossibility). Once an event has become and passed, it is immutable. Whereas when we think of the stage set for some possible event yet to come, whatever may be our expectation as to what is going to happen, and whatever our preparations are in anticipation of the particular happening which we calculate will suit our purposes, we know that it is possible that the expectation may not be fulfilled in the way we desired. In contrast to this open-minded attitude to the future, when we regard the world as a physical system obeying dynamical laws, we seem forced to argue in the following way. In order to determine the whole path of a particle moving in a given field of force by means of Newton's laws, all that we require to know is its position and its velocity at one instant of time. Similarly for a system of particles subject to any constraints which are clearly specified, provided that we have the mathematical ability, we should, from the specification of the system now as to velocities and configuration, be able in general to calculate the whole future motion. Now, we think of the whole world as a dynamical system of the kind just referred to, and consequently conclude that since Newton's laws are 'correct', the present dynamical state of the world determines or causes the whole future of the system. We have introduced Newton's laws in this connection because it was in this way that the conception of determinism found its way into physical thought, and also because some men have considered it important to show that quantum mechanics has liberated us from this kind of determinism.

What is the logical elucidation of this situation? We shall leave on one side, as essentially irrelevant, questions arising from limits to measurement and shall not ask what sense, if any, can be given to treating the whole world as a dynamical system. In the first place, then, let it be pointed out that the dynamical connection between the position and velocity of a particle now and its future motion determined by Newton's laws can be expressed in quite another way, namely, that in which the distant future appears to determine the less distant future. For, if we are given two points on the path of the particle and know the Lagrangian function for the motion in question, we can calculate the whole path by means of Hamilton's Principle of Least Action. This result led Hertz (*The Principles of Mechanics*, p. 23) to object to the principle in that it makes "the present motion dependent on consequences which can only exhibit themselves in the future". Now the form of the motion represented by a trajectory is a spatial one, and the equations describing it can be just as well written in the form of the variational principle as in the form of differential equations. Constants of integration are required in the latter and terminal conditions in the former. In the equations of the trajectory, there is logical connection between the initial motion and the subsequent motion, and this is as much an effect of the future motion on the present as *vice versa*, for connection between the parts of a symbol or idea, being a logical connection, is not in time at all in the manner that the particle connects the places through which it passes. It has no sense, however, to say that an actual motion now is controlled by the future motion which has not yet become, for we cannot settle the experimentally correct description of the system until the future motion has become.

We have to say that determinate connection of motions exists only in our representation. If we use Newton's laws, the connection has one form; if we use a different law, the connection has a different form, and is no less determined than in the former. Which of these forms agrees with facts is not determined by the laws at all. The Newtonian form has been successfully used in its particular way in the past, but it is only habit that leads us to expect that the future will be like the past; we may have—in fact we have had—to change the method of representation. In physics whenever we adopt a particular method of representing nature, we give ourselves up to the system of possibilities which it allows for the description of facts. Any state of affairs which can be described in ordinary language but which does not agree with the predictions of the relevant laws of physics is an *impossible* state of affairs relative to our method of representation; it is not one of the possibilities allowed in the method. When on account of experimental observation we wish to establish in the theory the possibility of what we have observed, we have to modify the method of representation, and once we have done so we have given a possible place for the new phenomenon. All this is a complicated matter, because although here we have treated of one method of representation only, in fact physics employs many methods, and the possibilities of describing nature vary with the method employed. In consequence the logical nature of causality tends to become hidden.

Now it is perfectly clear that the representation of nature by means of quantum mechanics has a different system of possibilities from that by means of classical mechanics, and in particular it does not allow the description of the motion of a particle with any degree of experimental refinement we care to imagine in the way classical dynamics does. Nevertheless, we are tempted to represent motions with a fineness surpassing what is allowed by the uncertainty principle of Heisenberg, and thereby represent to ourselves, as distinct possibilities of description by quantum mechanics, states of affairs which are in fact no more distinct possibilities than are the differing numbers given as the result of an experiment to measure the capacity of a condenser by means of Maxwell's method of repeated charge and discharge in a bridge, when these numbers all lie within the possible error of measurement. If the numbers did represent distinct physical possibilities we should be compelled to look for the causes of the various differences. The difficulty with quantum mechanics is this, that we make the kind of pictures we have been accustomed to make in classical dynamics—our method of representing them on a diagram allows them—and in order to introduce the limitations properly belonging to quantum mechanics we have to apply a rule in interpreting the picture. This rule governs the degree of fineness of the picture, and it is very easy to point to the possibilities allowed in the picture without the rule, and to ask questions as to the cause of differences of motion clearly represented in the picture. Suppose now that we choose a method of representing which will keep the rule constantly before us—we shall discuss such a method later—then we shall no longer be able to describe as distinct possibilities of motion any differences of finer grain than are allowed by quantum mechanics; we shall not be tempted to ask what are the causes of the differences which we were able to represent in the former picture. In this way the causality of the classical method of representation disappears from our description of nature. In its place we substitute the causality of the new method. This should have been obvious to us whenever we have reflected on the fact that theories employing quantum mechanics have been successfully applied to describe phenomena. The scattering of electrons or helium atoms by a crystal has a definite angular distribution which quantum mechanics enables us to predict correctly—if it does not, we have to look for a cause to explain the discrepancy. Quantum mechanics has substituted the causality of its laws for the causality of the laws of classical mechanics. Consequently the new theories of physics will not do for us what not a few men have wanted them to do, namely, to get rid of determinism. It is true that by insisting that classical mechanics be given up in the description of atomic processes, modern physics has removed the kind of connection typified by that between the path of a particle and its initial position and velocity, but it establishes instead the determinate connections which are required by its own methods of calculation. All that happens is that the calculations belong to different systems and different consequences follow. But the essence of determination, namely, the necessary logical connection between the data and the properly calculated prediction of the theory, is present just as in the old law. It has no more sense to say to-day that nature is not determined to obey the law of causality than a century ago it had sense to say that it is so determined.

The philosophical importance attached to the word 'determinism' arises from its being regarded as the alternative possibility to 'free will'. This is a muddle due to mixing the two uses of the words 'must' and 'necessary'. We are always tempted to confuse logical necessity which applies only to the correct use of symbolism and has nothing whatever to do with what is the case, with economic necessity, for example, according to which a man feels compelled to act in the way he does in order to ensure some satisfaction of his economic needs. Now this man may choose, for some religious reason for instance, not to act in accordance with his immediate economic interests. That there is no logical necessity in this is already shown by language itself which allows the choice. On the other hand, logical necessity is exhibited only by the absence of choice as to the result of calculation by a given rule. A man who wishes to play the piano correctly from a given score has no choice in the notes he plays; the score dictates. But on the other hand he might have chosen to play another piece of music; and there is no necessity in any case that he will in fact play the chosen score correctly.

In connection with what has just been expressed concerning determinism and free will, it is perhaps of value to refer to another famous dilemma, namely, the opposition of mechanism and vitalism in biology. We have to ask what the terms 'mechanism' and 'vitalism' mean. The former certainly refers to the application of the symbolic methods of physical science in all processes describing nature. Whatever descriptions we make of fact, they must be made in terms of some theory and of a method of representing nature. What constitutes the mechanism in our theory is the system of logical connections of the ideas appearing in the theory: mechanism in the sense of necessary connection properly belongs to theory only. Hence the so-called vitalistic hypothesis must have this in common with the other, if it is to rank as a theory at all and not as an abdication of the attempt to describe what happens. Biology can have—does have—its own special symbolic methods which are external to or added to the methods of physics and chemistry. But these special methods have to be used in the same way that symbolism is used in physics. We make pictures of fact and have a proper experimental method for testing their agreement with fact. The notion that physics can explain biological theories in the way that it is now attempting to explain chemistry is of the nature of a plan of campaign in science, whose appropriateness is to be decided by considering its usefulness, not by asserting its logical inevitability. Vitalism, however, cannot be maintained as the basis for biology if by vitalism we mean opposition to the mechanism of representation by determinate connections.

Writers who have wished to see in Heisenberg's uncertainty principle an escape from determinism in the description of atomic events, have found themselves in a predicament when faced with the well-established laws of classical molar physics, in which phenomena on a scale large compared with atomic magnitudes are involved. Accordingly they have taken refuge in the theory that all the molar laws are statistical laws which hold with great accuracy because the fluctuations from average behaviour are negligibly small on account of the large number of individual atoms taking part—whether of matter, of light or of electricity. The point of view is just that encountered in bringing together the molar laws of fluids and the kinetic theory of matter, except in one respect, that in the elementary physics textbook each molecule is endowed with mass and motion in the ordinary way for the purpose of explaining the mechanical properties of fluids, whereas according to the new view the mass and motion of the molecules are to be treated by dynamical laws which are equivalent to the classical ones only for gross phenomena. But it is the classical laws in terms of which we measure dynamical quantities in the laboratory by means of apparatus we can see and handle. The theorist who wishes to present physics according to the scheme that atomic and molecular systems behave only statistically in a regular way, may define mass and other dynamical quantities in whatever way he likes, but he has still to answer the question as to how these quantities are to be measured. Here he has to refer to apparatus to be treated in the way we have always treated it. The statistical theory does not destroy our common-sense view of the things we handle; it must be advanced for the purpose of explaining that view.

To state that the laws of classical physics are statistical does not undermine the determinateness of descriptions by means of them, for we have to keep in mind, as always in statistical mechanics, that we cannot expect a statistical law to apply with greater precision and detail than is allowed by the fluctuations in the physical magnitudes connected by means of the law. So far as the statistical law is concerned there is no cause for the affects we call fluctuations; they are accidents. If we wish to give a cause we shall have to use a method of representation which permits pictures of finer grain. For instance in atomic physics we always look for the cause of a change in the energy and momentum of an atom or electron and so on. On the other hand, to the question, why does one atom absorb light rather than another on which the light is incident? we can give no answer so long as we do not represent the atoms in different situations. The logic of this matter is no different from that of the tossed penny which shows head on one occasion and tail the next: if we are unable to solve the dynamical problem under prescribed conditions of tossing we cannot give a cause for the observed result. In any series of n tosses we can imagine 2^n possible results, but we have no rule by which to select on any given occasion any one of these possibilities as our expectation of what will be found in actual fact. So long as this rule is absent so long also is a cause lacking.

Let us now return to consider a method of representing motion which keeps before us a rule governing the fineness of grain of the pictures of fact made by means of it.

Imagine a chessboard extended indefinitely parallel to two

adjacent sides of the board which we take as axes of reference. Any square on the plane can be selected by giving two integers which correspond to the x, y coordinates of the corner of the square nearest to the origin of reference. If we use this system to name places on the plane there is no place between two adjacent squares, for there is no integer between n and $(n+1)$, and according to our system of marking places the possibilities of existence on the plane of the board are given by pairs of integers only. Now when a physicist is surveying positions on a plane relative to a system of rectangular axes by means of a metre stick, the coordinates he obtains when he applies the rule that only those figures are to be given which are significant above the possibility of error in measurement, he obtains pairs of numbers which are integral multiples of a common unit of length, say 0·1 mm. He has the choice of representing his measurements on the chessboard we have just discussed, or of using the geometry of the Euclidean plane according to which between any two points of any line on the plane there is the infinite possibility of finding other points. With the latter method, however, in comparing the coordinates of any point of the plane of his representation with the coordinates he can survey, he has to use the rule that any two points in the representation whose x-coordinates differ by less than 0·1 mm. and whose y-coordinates differ by less than 0·1 mm. do not represent places on the surveyed plane which can be distinguished with the apparatus at his disposal. Now in actual practice one does not use the chessboard or its three-dimensional analogue in physics because one would have to change the method of representation every time that the chosen type of coordinate system is changed, and whenever one avails oneself of a more precise method of measurement. That is, one chooses the usual continuous method of representation in preference to any of the group of possible discontinuous ones. It is much more convenient to vary only the rule for interpreting the picture than to alter the geometry of the space used in the representation.

Suppose now that we wish to study the translational motion of a body whose shape is never exactly the same, and that for practical reasons we are not able to follow the motion of the parts of the body and therefore of its centre of mass. We should not then be able to follow any definite point of the body and represent the motion of the body as the motion of a point, without a rule of interpretation; we should have to represent it as the motion of a smudge on our map, and this smudge remains of roughly the same extent throughout the motion. If we think of the motion of the body as the motion of a point we shall have to say there is an uncertainty in the position of the point at any given time. Suppose that we employ the chessboard method of representing this state of affairs, then to the question where is the body at an instant of time between t_1 and t_2 at which times it occupies the positions (m, n) and $(m+1, n)$, the answer is 'nowhere'. The method of representation we have adopted does not allow the possibility of a place between two adjacent squares. The motion of the body in the x-direction is discontinuous when we look at it in this way.

Now consider the motion of the point marked by a spot of light projected on a screen from a cinema film. Some twenty discrete pictures are shown per second, and between each pair of successive pictures there is a brief interval of darkness. When we look at the moving spot of light on the screen, its motion appears continuous, that is, we can represent its motion along its path by $x=f(t)$, $y=g(t)$, where f and g are continuous functions. Yet if we adopt a sufficiently refined method of observation, we find that we can specify some times when the spot is not moving and others when there is no spot of light on the screen at all. The motion of the spot thus appears to be discontinuous when examined by refined observation. On the other hand, we could still preserve a continuous representation by altering the form of $f(t)$ and $g(t)$ so as to allow the spot of light to remain fixed on the screen during the short intervals during which it is visible and to postulate any kind of continuous motion that we care to imagine in the dark intervals, provided that the initial and final positions for each such interval are correctly chosen and that the initial and final velocities are zero. In order to prevent the possibility of testing the law which we have proposed to govern the dependence of x and y on t, we have to add another hypothesis which amounts to denying the possibility of observing the spot during the dark intervals. These two methods—continuous and discontinuous—are equally allowable methods of representing the motion, but one is likely to regard the latter as the more appropriate and less artificial method.

The two phenomena just considered bring before us in a direct manner the possibility of discontinuous methods of representing motion, and in one the discontinuity applied to space and in the other to time. Secondly, the use of a discontinuous representation was suggested by the appearance of our picture when a continuous representation was made, and this is to be distinguished from the discontinuous representation which one introduces to exhibit the limits of fineness which are known to belong to the method of measurement adopted, as, for example, in the crude survey of a plane with a metre stick or in a crude temporal description of any process with the aid of an ordinary clock. In the latter case our pictures have a fineness of grain depending on the methods of measurement, in the former they have a discernible grain corresponding to the process which is being described. If the grain of fineness determined by the method of measurement is coarser than that belonging to the process, the grain of the latter cannot be represented at all. Let us leave on one side any restrictions on the fineness of the measurements, and choose for the purpose of representing the measurements a continuous space and time. Any point of space and instant of time represents a unique possibility for an event, provided that the nature of the event itself does not require a finite interval of time or a finite region of space in which to happen. On the other hand, if on account of the character of the event we cannot do better than represent it on our map as a smudge, which is also extended in time, then any two point-instants lying within a region of space-time corresponding to the smudge do not represent distinct possibilities at all.

For simplicity consider the motion in one direction of space, say parallel to x, no reference being made in our specification to y and z. We can use our chessboard to represent this motion, the direction of one edge of the board representing distance travelled parallel to x, and that of the adjacent edge representing time interval from an arbitrary origin of time. The sides of the squares of the board now represent in one direction the x-extension of the smudge in space, while in the other they represent the inherent uncertainty in timing the moving body. Let us denote by Δx and Δt the spatial extension and temporal duration represented by one square on the chessboard. In order to determine these magnitudes some experimental information will be required, and it is quite reasonable to assume that the smudge has an indefinite boundary which prevents us from fixing Δx and Δt more accurately than $\Delta^2 x$ and $\Delta^2 t$. Suppose that we are to measure the velocity of the moving body. For an accurate measurement we should time the body over a distance X great compared with Δx and time T great compared with Δt. In our representation, $X = m\Delta x$ and $T = n\Delta t$, where (m, n) are the integers naming the square on the board which represents the final position of the body relative to axes passing through the square representing the initial position. Now we are uncertain as to the exact size to give the squares of the chessboard, so let us consider another possible representation in which the squares are taken to represent $(\Delta x + \Delta^2 x)$ by $(\Delta t + \Delta^2 t)$. If we draw one system of squares on the top of the other we see clearly that the number of possible places on the plane where we can represent the appearance of the body is very much reduced, for whenever a square of one system overlaps those of the other, the possibility asserted by the one is contradicted by the other, and only when a square of one board does not overlap those of the other is there a clear possibility of representing the existence of the body in space and time. Now these possibilities must agree with the observed initial and final positions of the body whose motion is being studied. Consequently we have

$$X = m\Delta x = (m - r)(\Delta x + \Delta^2 x), \quad T = n\Delta t = (n - s)(\Delta t + \Delta^2 t),$$

where r and s are integers. In order to proceed further, we must consider a particular kinematic law of motion—so far we have dealt only with a method of representing the motion. We shall deal with that law which corresponds to uniform velocity in a continuous representation. That is, we have $r=s$. In order that the measurement of velocity may not have an excessive uncertainty we have to choose m and n as fairly large integers, and then we have the approximate relation

$$v=\frac{X}{T}=\frac{(\Delta x)^2}{\Delta^2 x}\Big/\frac{(\Delta t)^2}{\Delta^2 t}=\frac{(\Delta x)^2\,\Delta^2 t}{(\Delta t)^2\,\Delta^2 x}.$$

It only remains to point out that this result is exactly the same as one obtains for the group velocity of waves of wave-length $\lambda=\Delta x$ and period $\tau=\Delta t$, $\Delta^2 x$ and $\Delta^2 t$ being the range of wave-length and period covered by the group. The uncertainty in the value of v is of the order Δv given by $X\Delta v=v\Delta x$, which is just the result obtained by applying Heisenberg's uncertainty principle. Indeed, there is a splendid opportunity here to give the appearance of an *a priori* derivation of the uncertainty principle. In the general case it can be shown quite easily that $\left(\frac{1}{\Delta x},\frac{1}{\Delta y},\frac{1}{\Delta z},\frac{1}{c\Delta t}\right)$ transforms like a four-vector, just as the energy and momentum. These two four-vectors, one says, must be related by a constant of proportionality because they are all we know about the particle whose motion is being described. Take h as this constant of proportionality, and at once we obtain the uncertainty principle of Heisenberg. I have thought it worth while to draw attention to the possibility of this argument, because it is typical of what some philosophers attempt to do with physics. These writers try to make one believe that there is an *a priori* proof of physical laws. Now it is true that one can exhibit the logical relation of any particular physical law to considerations of a general character, but one has not the slightest justification for supposing that the relevance of the particular law for physics is established on other than the basis of experimental usefulness. Indeed, it is highly doubtful that anyone would arrive at the particular notions which have found physical application if one had not been guided to them by experimental study of the facts.

It must be pointed out that the method of representation which has just been discussed serves a quite different purpose. It is not produced to give the impression that there is no logical alternative to Heisenberg's uncertainty principle as the basis of describing atomic processes, but in order to show that a method of representation exists which automatically abolishes the possibility of asking questions of the type which are often referred to as the paradoxes of modern physics. The method of representation adopted here is logically equivalent to the method which physicists adopt when they speak of the waves associated with a particle, but in one respect it is clearer in that we do not assume a wave process and we do not have the possibility of representing in the motion of the particle more than is required to describe the facts. It shows what is meant by discontinuous motion, and that the motion of the particles of atomic physics *can* be thought of in this way. It also exhibits very clearly how discussions about failure of causality are quite mistaken in their origin.

Before leaving the discussion of this way of looking at the motion of electrons, etc., I wish to make quite clear that I do not have in mind at all the possibility of using this method of representation as a desirable one for modern theoretical physics. There is no logical objection against the use of imagined wave processes to do what can be done without them, provided that we keep clearly in mind what we are doing when we represent nature by means of them. The virtue of the particular method mentioned here is that it brings out clearly a certain aspect of the logic of representing nature by quantum mechanics and does so in a manner which disposes immediately of philosophical questions which would never have been asked in any case if the part played by rules of interpretation in symbolism had been clearly understood. At the risk of boring the reader with references to maps and projection, let us recall the orthogonal representation of the earth on a plane. The picture obtained is a double one. It is necessary to have a rule for interpreting the map so that the projections of points in the northern and southern hemispheres are not confused with each other. This rule of interpretation belongs to the particular method of projection used; there is no need of this rule when we employ a spherical or cylindrical map of the earth, although in the case of the cylinder a special rule is necessary to interpret the map at the poles even when we are not dealing with metrical relations on the map.

Let us try to state briefly once again the attitude which has been taken up here in thinking of mechanism.

Mechanism in the natural sciences is the mechanism of our theories and is the same as the mechanism of calculation. The mechanism of theories means the determinateness of the descriptions by means of them. There are no other descriptions than those that are mechanistic in this sense. Apparent indeterminateness of description arises when we represent the pictures given according to our theory by a method which allows a finer grain of representation than the theory actually attempts to provide, and when, in addition, we forget this matter of fineness of grain in interpreting the picture. A crude map is perfectly definite in what it represents if we have in mind the degree of fineness of representation intended when the map was made. This consideration applies to all knowledge. Our thoughts need not be complicated to be clear in certain important respects. In physics we can think in no other way than this. We propose a law to describe a particular kind of process of which we have experience. If it is a good hypothesis the calculations by means of the law agree with the facts. If the process does not go according to the law, we propose another or invent an additional hypothesis to square accounts, and so on. Because we use a theory to describe processes we are tempted to transform the logical necessity of the descriptions according to the theory into a physical necessity on things to behave as they do. But there is no physical necessity—only logical necessity. Causality, far from being a law of nature open to experimental investigation, corresponds only to the correct use of symbols—to the correct application of theory. Differences that can be represented, account being taken of the symbolism of experimental measuring methods, must be described by different rules.

In giving a physical explanation of the rule we give the cause of the difference in question.

Since we have dealt in detail with some of the questions which have become prominent in modern physics, in leaving consideration of them, it is not out of place to draw attention to a matter that is often overlooked by theoretical physicists, for by the omission some people may have been misled who are not familiar with the actual facts of the laboratory. In expounding the principles of physical measurement taking into account Heisenberg's uncertainty principle one discusses their application to imagined experiments, such, for instance, as the interference of electrons allowed to pass through a system of diaphragms. Now it is quite true that such an arrangement is, on account of its simplicity, much better suited in principle than the actual experiments which have been performed, to bring to light the logical point one is striving to make clear. But this granted, one should add that the imagined experiment never has been performed, and, so far as an experimental physicist can see, has hardly the slightest chance of ever being attempted, because of the extreme difficulty in making the apparatus sufficiently refined to show the effect in question. This is a significant statement about the world.

CHAPTER V

THE LOGIC OF SUBSTANCE AND MOTION

IF one wished to explain the idea of a substance to a child one might show him, for example, the lumber in a carpenter's yard, let him watch the carpenter at work making tables, chairs and other articles with which the child is familiar, and then hope that he understands that all these things are made of wood, the substance of wooden things. The particular form—the table, the chair, the log—is accidental, it might have been otherwise; whereas the substance of wooden things is wood and is unchangeable; the same stuff is in the log as in the chair and the table. Now this does not mean that tables and chairs cannot be made of other materials, nor does it suggest that there is one kind of wood only. What is implied is that the substance of wooden things is wood as opposed to stone or iron, and that a table made of iron is not a wooden thing. This is what we think of substance in everyday life, and it is this notion as it appears in physics which we have to analyse. The table is made of something which exists by itself, whether it is fashioned in the form of a table or not. This something occupies space and cannot be at the same place as something else at the same time. If we wish to know any more about it we very soon enter the study of physics and chemistry, which more likely than not will be irrelevant for our logical inquiry.

The adjective 'substantial' has a variety of meanings in common use; since these seem to have infected the word 'substance' and served to render unclear the meaning of so important a term, it is desirable to examine them. Here are three: 'actually existing', 'real', 'solid'. Of these the third need not detain us long. In physics, solids, liquids and gases are all substances, and the educated man understands this implication in ordinary language. The term which is used to denote any or all of these is 'matter'. Association of substantiality with reality and actual existence, on the other hand, has been responsible for some confusion in physical thought. In the first place, if substantiality is identified with reality, then it is not peculiar to substances as properly understood in physics, that is, to matter. The optical image of a metre stick which we see through a lens does not give rise to the tactual experience which we associate with the stick itself, and in this sense is insubstantial; but it is a real phenomenon, and the theory of the formation of such images is a well-established part of physics. On this ground optical images should be regarded as substantial, but such use of the term is surely an offence against the English language. The optical image is real in the sense that it is a recognised physical phenomenon, but the mere appearance of the stick where the image is, does not entitle us to infer a 'real' stick in that place, that is, a stick which has the usual properties of matter. For this reason the image of the stick is insubstantial; the apparent stick in the position of the image does not actually or really exist. The distinction between the two sticks is that which is epitomised in the phrase of common speech—'substance as opposed to shadow'. Paradoxically enough, however, the shadow itself must have substance—otherwise we could not see it. If we see the shadow on the ground, those places on the ground where there is shadow must differ from those where there is no shadow; that which distinguishes shadow from not-shadow is a substance, though of course not the substance of the object which casts the shadow. For the same reason, the optical image has substance; it is not made of matter, it is true, but something must mark the points of space or of our visual field where there is an image in order to distinguish them from those points where there is no image.

In mathematical physics substance marks places in space, and according as the particular problem which happens to be under consideration requires it or not, is regarded as extended in space or not. In classical physics it always persists in time. But it has another important property, namely, the law according to which its spatial coordinates are to be transformed. The only transformations which are permitted are those which belong to the group of continuous motions. Substance cannot disappear in one place and suddenly reappear in another distant place. This rule is made completely definite for a continuous medium, by means of the concept 'quantity of substance'. In any motion the quantity of substance which is contained in any closed surface S is equal to that contained by S', where S' is the surface into which S is transformed by the motion. It seems, then, that mathematical physicists have merely taken over into scientific theory the common notion about substance and that the idea of the conservation of quantity of matter, which appears to have been added from the fund of experimental discovery in the late eighteenth and early nineteenth centuries, is nothing but the principle to which man has always trusted in dealing with his environment, and which is part of the logic of common experience. It is what makes us refuse to accept the production of something out of nothing by a conjuror as anything but a trick or illusion except that in physical science we require a balance to settle the matter. On the other hand, one might argue, since substance is inseparable from the possibility of its motion, and since something must remain unchanged during the motion, some sort of conservation theorem is necessary; and that what mathematical physicists have done is to adopt the particular theorems which are in agreement with experimental knowledge. Now, does a conservation theorem secure the kind of invariance with respect to motion that is required? The answer to this question is of great importance, for if it does not, then the use of the word 'substance' to denote that which is conserved during motion is a special technical one and should therefore be clearly distinguished from the word which has interested the philosophers.

In the game called 'billiards' there are two white balls, which, as accurately as the maker has been able to secure, are identical in all save one particular, namely, that one is marked with a small black spot, whereas the other is not. This spot alone serves to distinguish one ball from the other. Whatever the motion of the balls, the spot is carried along by the same ball all the time; if we wished to test this however, we should have to use some other means of distinguishing the balls and then see whether or not the two methods of identification agreed. If they did not agree, we should be at once engaged in investigating the peculiar behaviour of the spot on the billiard ball, or a case of mistaken identity arising from some other cause. Whatever happens, we must have some means of distinguishing the balls which is independent of their motion. This is what remains the same, independent of the accidents of the history of cannons and losing hazards. This is all that need remain unchanged in the motion of substance—the possibility of identifying it. Provided that in the process we do not remove this possibility, we can subject a piece of matter to extremely drastic treatment, and still talk of it as the same. The cigarette which was lit five minutes ago is now half consumed, it is still referred to as the same cigarette. We see therefore that the permanence of substance is the continuing possibility of recognising it: what remains the same is its name, which, of course, is a symbol. We express the unchangeability of any thing with respect to motion and other transformations by using the same word by which to refer to it. That word may be correctly or incorrectly used, to settle which we must consider the meaning of the word, and compare all that the meaning involves with what is the case; for we do not create what is unchangeable merely by reiterating a word. The conservation of quantity of substance is clearly added to the invariance of the properties which allow us to use the same name.

When we represent a state of affairs, we do so by making signs, marks on paper as in a picture, models, writing, speech, or gesture by touch or sight. All these methods present to us objects of sense. As matter is said to exist, so these signs exist, and by their existence now present one particular possibility as opposed to any other which might have been presented as the state of affairs intended. In the picture we see clearly that the marks on the paper distinguish the points of the plane which are marked from those which are not. Now, instead of actually drawing the picture we could write down the coordinates of the points which are to be marked and how they are to be marked, as in telegraphing a photograph, and just as a chess game can be described (and played) without the actual pieces in position on the board, merely by signalling the moves in the well-known shorthand associated with the game. Naturally the possibility of rendering the picture by means of a catalogue of coordinates applies only to a picture of finite mesh or grain; whenever the picture is a continuous one we should have to give laws by which continuous sets of points are to be selected for marking, such as equations of curves. With this way of presenting the picture, its description in words bears a close connection; the words can be translated into the laws according to which the form of the picture is to be drawn. But it is not to be assumed that a word is necessarily equivalent to a mathematical equation—ask how the word 'horse', for instance, functions in language. The words do in some cases name laws or define them, and in this respect they play the rôle of the pencil marks in the drawn picture. These marks are the substance of the picture, and so long as we are interested only in geometrical form might be made in a wide variety of ways as by ink, chalk, graphite and so on, without in any way interfering with the representation. All that is required of substance here is to mark loci in a geometrical space, and so far as the representation is concerned, all other properties of the actual matter used (as for instance, the colour of the ink) do not count. Whereas, if the picture were a coloured one, then each coloured mark indicates a point not merely of geometrical space, but also of colour space. Of course, coloured marks need not be used to represent colour, and in fact are not used in the written description, where words or other written signs serve this purpose.

In all representation, substance plays the essential part of marking selected places in spaces, and as in the pictures we make so in the reality they are intended to represent. Substance is independent of the particular picture according to the selected method of representation; it is the same in whatever place it is. A changing state of affairs is represented by the motion of substance in the space of the picture and by transformation of one configuration into another. In translating one kind of representation into another, we carry over the motions and transformations by some law of projection, which very often is, but need not always be, that according to which the same transformations occur in the two representations; for example, the representation of motions on a spherical surface, by means of motions on a plane, requires the rule for projecting the sphere on the plane. But this does not mean that uniform motion on the sphere is represented by uniform motion on the plane. Configurations marked out by substance in one representation are translated or projected into the corresponding configurations marked by substance in the other; but what the substance is in the first representation is not represented in the second; only that there is substance where it is in one, shows, when the translation has been made correctly, that there is substance in the corresponding places in the other.

Substance, then, stands in a universal logical relation to representation. This is what gives the concept its philosophical importance, but instead of making the proper logical analysis, philosophers have usually given it a mystic or occult significance by referring to it as 'the permanent, unchangeable reality behind the ever-varying phenomena', or in some equivalent phrase, and by concerning themselves with the so-called ontological problem, which is no more a problem than any other logical puzzle. The subtle distinction which has been drawn between an image 'in the mind' and one presented by visible signs rests on the supposed certain existence of the substance of the former as opposed to the hypothetical existence of the matter in the case of the latter. Now, as a matter of fact, we do settle whether or not the signs exist whenever we are confronted with the possibility of hallucination, or other source of error, in the same way that we decide in favour of any hypothesis. For instance, consider the sentence 'there is an oak tree at the north-west corner of my garden'. Are we to look on this sentence as a proposition or as the expression of a hypothesis, that is, a rule for making propositions about the garden and the tree? We cannot exhaust the possibilities of a hypothesis—consider the innumerable appearances presented by the tree viewed from different places, in varying states of the sky and so on—there always remains the chance that although all those propositions that we have tested are true, one or more that have not been tested might not be true, in which case the original hypothesis would have to be given up, or if we are unwilling to do this, a cause must be given to explain the observed appearance. For example, I may look from a window, notice the absence of the tree, and on closer inspection find a sawn stump in the place where the tree used to stand, whereupon I reflect that someone must have cut down the tree and removed it. When, on the other hand, we regard the sentence about the oak tree in the garden as a proposition in ordinary language, we decide its truth or falsehood without obviously examining the matter in the careful manner we adopt in dealing with a scientific hypothesis. If we decide that the proposition is true, then we also say that the tree exists, and this is one perfectly correct use of the word 'exist'. Consider the physicist in his laboratory; does he ask questions as to the existence in this sense of his apparatus, unless he is unable to see or touch it? Yet some writers have tried to say that physics is not concerned with the existence of the objects of which it treats.

The system 'exist : not exist' is quite properly applied to anything that can be used as a sign in ordinary language, or can be described in ordinary language. Language leaves a place for signs. There is no sense in questioning the application of this system to trees, birds, stones and men, and at the same time accepting it as applied to the images of these things presented in our visual field. When I wish to say 'it is inconceivable that experiences other than my own are real', or 'only the objects immediately presented by my senses really exist', I am proposing a notation different from that of ordinary language, and the justification which some men have wished to give this different notation is *senseless*. It is not proposed here to follow the philosophical elucidation of solipsism, which is not to be disposed of in the easy way adopted by common-sense philosophers in attempting to dismiss it; the elucidation is a long process requiring detailed examination of a number of linguistic expressions.

To say that an idea exists can mean either that the word or words we have used to name or define it have a meaning, that is, stand for a possibility of language, or it can mean that an idea exists in our thought. Are these uses of the word 'exist' the same?

The question whether matter really exists has always appeared to be a serious conundrum for physics. By means of it philosophers have attacked physical theory with a peculiar disregard for the practical use of the subject, and without producing any abatement of the enthusiasm in which physicists—and other scientists—have pursued their investigations. At the same time not a few physicists, by slavish adherence to a naïve materialism, have imposed restrictions on their method of thought about physics, uncalled for by the logic of the facts, although they have defended their views with conviction, and have therefore found it impossible to follow with intellectual comfort any revolutionary change that has taken place in the subject. Such a revolutionary change is always in the method of representing phenomena. In its asserting the existence of a world 'outside of and independent of ourselves', materialism tries to express in words what the very application of language itself shows. In its insisting that we can know something about the world without knowing all, it attempts to express the possibility of language. And so on; in all those respects in which materialism has been of service to science by liberating it from other philosophical theories, it has been a case of taking over into what was apparently a theory, matters which ordinary language itself looks after when it is properly used. However, since materialists do not recognise this and since symbolism is ignored by them, their position has been easily assailed by philosophers whose theories never could be of service to science.

Metaphysicians used to be concerned with the elementary or ultimate substance or substances and the tradition has not yet died out. For many people still, physicists are probing till they come to something beyond which they cannot go, something that has metaphysical importance; whereas the terminus of all investigation is an adequate theory to describe what happens, and the importance of the work is not intrinsic but lies in its usefulness.

Ordinary matter is composed by chemical synthesis of substances which are elements with respect to chemical analysis. By means of other physical processes the chemical elements are analysed into atomic nuclei and electrons, and so on. Thus there is a series of methods of representing ordinary matter; in any one member of the series the substance of the preceding one is given internal structure by means of which to account for its properties. Such a series need have no last member, but we can stop anywhere in the series that we find convenient. For example, the organic chemist is rarely concerned with the physicist's analysis of the chemical elements; chemical symbolism has the possibility of representing most of the facts that he has to deal with, consequently he uses it and justifies doing so by the purpose he has in mind. On the other hand, whenever the chemist attempts to construct a theory which will explain the laws of chemistry, he has to employ physical symbolism, or produce a substitute for it—a responsibility which is not always faced. The modern attempt to account for the formation of molecules by means of quantum mechanics and the models of atoms which serve to explain the spectra of the elements, must be regarded at present as a means of testing the hypotheses of the theory, rather than as a means of predicting what compounds are possible for practical purposes.

We have seen that in any representation all that is required of substance is to mark places in the space of the representation. The properties of the substance are not represented; but they come into account in the making and recognition of signs. In physics and chemistry, however, the properties of substances do appear to be represented, just as they are described in ordinary language. A piece of gold has a shape, a colour, a hardness, a weight and so on. A different piece of gold would differ from the first in any or all of these respects, but not in the physical properties, density, spectrum, possibilities of crystalline structure, specific heat, melting-point, latent heat of fusion, electrical conductivity and so on, although a certain tolerance must be allowed for impurity, and for differences in elastic and thermal history when we make this assertion. The number which denotes the density of gold varies slightly from one determination to another, so that a range of possible values in the neighbourhood of a conventionally accepted mean represents gold as opposed to iron and other substances of different density, on the density scale. If the ranges of possible values for two substances overlap it is no longer possible to distinguish them by density measurements alone. Actually the density of every substance depends on temperature; consequently gold has to be represented by a curve on a density-temperature diagram, from which certain small deviations are allowed on the grounds already stated. There is no need here to enter into all the practical details which are of importance to physicists. That a substance has a particular density means that we mark the corresponding place in the density scale (or one-dimensional space). The system for naming places in this space, namely by numbers, differs from the system of words by which we name places in colour space, for the variable number can be used in calculations according to the laws of ordinary algebra, whereas the variable colour cannot.

To say that a thing has a certain property means that the thing occupies a place in the corresponding space. We cannot think of the thing apart from this possibility. A thing has geometrical form and has a place in the space of each property, so when we speak of the substance of which the thing is made we have in mind that the appropriate places in the property spaces are occupied while the geometrical form is left variable. We can therefore represent a substance by the proper point in the multi-dimensional space which is the product* of the property spaces. That a thing is made out of a particular substance requires that with each point of geometrical space there is associated the same point of the product space, but the

* Not 'product space' of group theory.

possibility that the thing might be otherwise is thought of by associating the whole product property space with each point of geometrical space, i.e. no place is marked in the property space. For each new property of a substance that we represent, an additional space is required, but the properties of the substance by means of which the particular place in that space is marked are still not represented. It should not therefore surprise us when we find it convenient to represent properties by special numbers like matrices or in spaces of which we can form no adequate three-dimensional model.

That we represent one property of a substance does not require that we represent any others. The space in which we represent a property is not the space in which we represent the process by which we recognise the property in question. Thus the space of the physical property 'mass' cannot be used to present the entire process of measuring a mass, but, of course, this process is what distinguishes a measurement of mass from a measurement of, say, specific heat. When we measure a particular mass by weighing in a balance, we have to count the standard 'weights' in one pan of the balance; the result of the count is the number which denotes the measured mass (leaving out of account any corrections that may have to be made); counting is part of the process of measurement, and this is all that is represented. In weighing we count 'weights' which have been compared with each other by means of the same operation of weighing. This involves that the numbers we obtain in weighing represent in a different way from the numbers which we should obtain in judging colours after having arranged to substitute numbers in place of the words 'red', 'blue' and so on. In the latter case the fact that the numbers are used to calculate according to the laws of arithmetic is irrelevant, in the former it is absolutely essential. The properties of substances which appear in physics are properties relative to the corresponding methods of measurement, and rules of calculation. For example, roughly speaking, a determination of density requires the measurement of a mass and of the corresponding volume, and the calculation of the ratio of the former to the latter.

The properties of a substance should be independent of the geometrical form of things composed of it and also of all other properties which depend on their extension in space. In actual practice, however, the result of the measurement of a so-called property does sometimes depend on these factors; it is then said that the property is so dependent, but this is merely a figure of speech. The properties of a substance cannot change unless we allow the same name to apply to more than one place in the combined space in which the properties are represented: it is then necessary to give a rule or law for selecting the places to which the same name may be given. On the other hand, the name of the substance may select a place in the space in which only some of the properties of the substance are represented; with respect to all other properties, the whole of the corresponding space is allowed; that is, in the space of all the properties the name of the substance selects not a point but a manifold of points, just as, in three-dimensional geometry, giving values to two coordinates only fixes a line on which the third coordinate may take any value whatever. For example, in dynamics, the colours of objects whose motion is described are not represented.

For the representation of the phenomena of the changing world, we choose a space. If we wished to represent a succession of coloured images, we should require two spaces at least, one to present colour, the other to present the temporal order of their appearance. Such a system has not the possibility of representing the shapes of the images: for this purpose another space of at least two dimensions is needed. (There is no shape in one dimension, only position and extension.) For every additional possibility of representation, the corresponding space is necessary. Ordinary language leaves a three-dimensional space for the description of the things we encounter when we move about. This space is used in physics, but the methods of surveying by which the relative positions of things are fixed have been selected for reasons that involve a vast amount of physical knowledge, and are therefore specialised in a somewhat arbitrary way. Nevertheless, what is in common between the estimation of distance by pacing and its measurement by means of rods or chains, namely, the repetition of an operation (taking into account 'right:left', 'forward:backward' and 'up:down') is all that is represented. Except in the description of static states of affairs, to this space is added the space in which time measurements are represented. Here again all that comes into account is counting the repetition of a process, together with the system 'before:after' which is incompletely represented in that if a succession of points in one direction of the space represents a possible succession of events, a succession of points in the opposite direction corresponds to no possible succession of events. The specification of processes which will serve for the measurement of time according to the conventions of experimental physics is an intricate one and rests on theoretical as well as practical considerations.

Unless there is a philosophical confusion to be elucidated it is out of place in dealing with the philosophy of physics to go into the practical details of actual conventions and procedures of measurement which are well described in technical works on experimental physics. Roughly speaking, philosophical problems do not lie there, because confused use of language renders one impotent in practice, and must therefore be got rid of. For example, if one is not clear about the definitions of the practical electromagnetic units as opposed to the international units, one cannot deal with situations in electrical practice where the distinction between the units makes a significant difference in the answer to a calculation or in the result of a measurement. Again, in making small experimental corrections in the laboratory one has to be clear as to the manner of applying them if one wishes to apply them at all. It is only when there is a doubt as to how a rule is to be applied that we feel the discomfort so familiar to the philosopher, and in practice one gets rid of this feeling when the doubt is removed by making the rule clear and unambiguous. It is otherwise when one begins to theorise about experimental physics—not in the sense of making theories which properly belong to physics, but in the sense of attempting to make generalisations based on simple analyses of the operations performed by experimenters in the laboratory. One is liable to overlook matters of practical importance in the attempt to force the description into a particular arbitrary form, chosen, more often than not, on account of some impressiveness or other character of the theory irrelevant for physics. Such an activity is usually concerned with nothing more than presenting our complicated science according to a particular scheme, in the belief that by means of it, the logical structure of the science is made clear. It is an attempt to impose a particular notation in which to express our thoughts about experimental physics and is akin to what philosophers have so often tried to do with ordinary language.

As a rule the representations of nature made in physical theory are simple compared with the experimental operations that have to be carried out in order to test whether or not they represent the facts correctly. This means that we represent mathematically only a little of the logical form of physical experience, and this gives physical theories their wide applicability. In this connection it is of value to compare with ordinary language what happens in physics. The use of symbolism is the same, but in physics we seem to make sharper distinctions and to draw sharp boundaries where there are none in ordinary language, and by elaborate experimental technique have provided ourselves with more refined means of testing the truth of the propositions we make. When we have learned to look on physics as a highly articulated language, we shall find difficulty in subscribing to the theory that physics proceeds by 'abstracting from nature'. Such a theory presupposes more complicated representations of facts than physics deals with, whereas in actual fact we proceed from simple representations to more complicated ones. One does not say that ordinary language works by a process of abstraction. In analysing the logic of representing motion we have to keep this in mind, otherwise we are tempted to make theories which cannot do other than mislead us.

Let us examine the device which we call a frame of reference. How does a set of three mutually rectangular axes represent on one occasion the rigid supports of an apparatus to determine the acceleration of a falling body and on another occasion an aeroplane in which certain physical measuring apparatus is carried? How can it be said that we abstract from nature in these cases when there may be no objects or boundaries of objects which remotely resemble in geometrical form the three mutually perpendicular lines? In any actual process of surveying one proceeds by first selecting a suitable base for measurement, represents this in the map, and then after surveying by measuring angles and chaining distances completes the map with the desired information so obtained. One usually draws an arrow on the map to represent the direction 'north' and writes names opposite the marks on the map. In order to use the map in practice, for convenience one 'sets' the map at a place represented in it by orienting it so that the arrow points north; when this is done the direction to any landmark coincides with the corresponding direction on the map. It is not necessary in using the map to start off where the surveyor did. Suppose now that one wishes to name a place represented in the map but which has no place-name; one usually gives the equivalent of cartesian coordinates relative to the left hand and lower edges of the rectangular border of the map as axes of coordinates. When we wish to find the place represented by these coordinates we do not proceed by looking for the axes of coordinates on the ground, we look for the nearest easily recognisable places, and starting from one of them use the map to determine a rule for finding the place which we desire to reach, by giving a starting point, a compass setting and a distance; and when after applying the rule we think that we have reached the correct place, we take some bearings to make sure that we have carried out the instructions of the rule correctly.

The procedure just described is a possible one but it is not unique. All sorts of devices are used in practice to find places that are represented in a map, depending on the particular conditions that prevail; the compass needle is of no use near the magnetic pole, the terrain may be so difficult that we cannot follow an instruction to measure the distance between two places by means of a chain and so on. What we do is to use the laws of geometry to choose a rule for finding the place which corresponds to the mark given by certain coordinates on the map: the particular rule chosen depends on practical considerations. Giving such a rule is the same thing logically as changing the origin of coordinates and employing, for instance, polar coordinates instead of cartesian. We use cartesian coordinates to name *places in the map*; the lines we draw on the map as axes of coordinates do not necessarily stand for any *things* in the reality the map is intended to represent. The laws of geometry tell us how to translate from one system to another for naming places on the map by means of coordinates, so that if we are given the coordinates in one system of reference we can find by calculation the coordinates in another system. For the moment we are considering only those transformations of coordinates which do not explicitly introduce time measurements. The particular system of coordinates which we use in naming places in the map can be replaced by any other system which has the proper logical multiplicity. On a plane two numbers are necessary, and in kinaesthetic space three. When we choose a frame of reference in experimental physics, we give rules by which any possible method of *naming* places on the map is to be applied in the actual process of *finding* the place represented in the map and conversely. Generally speaking certain configurations of objects are practically suitable for the purpose of connecting the map with our experience of what is represented in it, and we then speak of the axes of coordinates chosen in the map as being fixed to the objects in question. The process of fixing is not to be interpreted literally as gluing the map on a table for instance, but is to be understood as a definite rule for *applying* the representation which is made in the map.

Whenever a physical object whose position is subject to measurement at different times has to be represented by varying coordinates, we say that the body is moving. The kinematic law of its motion is expressed by the dependence on the time of the coordinates relative to the axes used. The representation of instants of time by means of points along a line has to be understood in terms of the methods of using the names given to the points by numbering them. We choose one instant of time as a datum which we call 'zero' in counting the time, but we could as well have adopted another origin of time. The transformation from one system of naming the instants of time to another is given by the geometry of the line. We fix the time scale to reality by making a particular point represent a particular observed event. We shall not discuss at all the elaborate processes by which one applies the representation of time when one has to use more than one clock, or when one wishes to employ astronomical time, any more than we shall attempt to consider any of the detailed practical considerations that enter into surveying spatial relations by means of different measuring devices. What these experimental procedures are is not represented in the map, but they must be understood and applied when the map is to be used in practice.

Whenever we have to deal with a moving body, the question arises as to the rule to be employed if we have to depend on it in applying the representation of our map. In translating names from one system to another, we make use of kinematics, which is the geometry of space and time representations. The simplest transformations which have to be considered are those called uniform translations and uniform rotations, and these, of course, are relative to the system of reference to which our map is fixed. All that is logically required to tell us how to translate the names from the one system to the other are the rules of transformation from one rectangular (or other) system of reference to another whose origin and axes of reference are named with respect to the first by equations which involve the time. When we consider the transformation from measurements of space and time by instruments on an aeroplane to those by instruments fixed to the ground, we apply the rules of kinematics, but we do not infer that the axes of reference necessarily represent any things in the state of affairs represented. These axes of reference are part of our symbolic machinery; they appear in a map or diagram, and with the aid of geometrical and kinematic laws we use them to calculate rules in which actual things are represented.

In the present discussion let us leave out of account the methods of representation used in the special theory of relativity. If we represent a moving body as a moving point, the motion in a system of representation in which we use cartesian coordinates will be expressed by equations in which x, y, z are given as determinate functions of the variable time t. If the system of coordinates be changed by mere displacement of the origin of coordinates, the directions of the axes being unaltered, the values of x, y, and z will be changed by an amount independent of the time, but the instantaneous velocity of the point will not be altered by such a change of axes. It appears then that when we describe the motion of a point by specifying, not the coordinates of position, but the components of velocity as functions of the time, the law will be the same for any system of axes which is obtained by mere parallel displacement. Such a kinematic specification is a law invariant with respect to this particular transformation of coordinates. From the mathematical point of view we know that the specification in question, being by differential equations, is indefinite with respect to the additive constants which appear on integration of the equations, and which are given values when we choose our axes of reference and when we are given the coordinates of the moving point at some particular time. On the other hand, if we wish to apply the law stated with respect to the system of axes A, to the representation by the system B which is itself moving relative to A, then we know that the kinematical law must be altered; the kinematical law which is invariant with respect to parallel displacement of the axes is not invariant with respect to uniform translation of them. Let us carry the process one stage further and notice that if the kinematical law of motion gives the acceleration of the moving body in its dependence on the time (but *not* on space) then the law is invariant for any parallel displacement and uniform translation but not for accelerated translation. Instead of looking on the kinematical laws of motion that appear in dynamics as being indefinite with regard to the initial position and velocity of the particle, we *can* regard them as indefinite with respect to the frame of reference to the extent of change of the origin of coordinates and of the uniform velocity of translation of the system of reference, but we cannot regard them as indefinite as regards accelerated relative motion of the frames. The fact that Newton's laws of motion involve the second differential coefficients of the coordinates with respect to the time, can, when the force is given as a function of the time *only*, be expressed in this way: that these laws imply a principle of relativity as to which transformations of coordinates are possible without introducing dynamical effects. So long as no velocities comparable with that of light are involved, this principle—the Newtonian principle of relativity—is in agreement with experience.

It has perhaps been worth while to draw attention to the way in which transformations of coordinates can come into consideration without any mention of an observer attached to each frame of reference. Observers have been introduced into discussions of relativity through the analogies in which one considers the physical phenomena from moving railway trains and from stationary platforms. But clearly the word 'observer' is irrelevant unless we have in mind someone equipped with measuring apparatus who actually does observe. Transformation of coordinates *need* not (although it may) introduce a hypothetical observer on every new system of axes we have to consider. This is a point of some logical importance, for one is tempted to require that the various supposed observers have to agree as to what they find as the laws of motion, when indeed there is no compulsion of this kind on observers in practice. The requirement that dynamical and physical laws shall be covariant with respect to certain if not all transformations of coordinates is a principle to be followed in devising a method of representing nature. It should be compared with the principle of the uniformity of nature.

Some writers on relativity have wanted to insist that physical objects should be invariant with respect to Lorentz transformation, for instance, and that this invariance is the essence of objectivity. There is in the theory of relativity, or any other physical theory, no opposition of objective and subjective, for the symbolism allows only the objective. Much of the so-called philosophy of relativity is irrelevant to understanding the theory, for it ignores the processes of representation which alone have to be explained. In fact the proper kind of explanation shows to what extent we understand the logical apparatus of non-relativity physics. Even if it should turn out that Einstein's general theory makes predictions found to disagree with observation, nevertheless we shall continue to think about physics with the added clarity which his work has given to our thinking.

In the description of atomic events by means of relativistic methods there is a difficulty to be clarified. It is the idea that one observes the same events from different systems of reference. If a light quantum turns up in the recording apparatus of one observer it is quite certain that it has not appeared in the apparatus of any other. How then can we say that two observers observe the same atomic event? So long as it is a question of merely finding the space-time coordinates which must be allotted to the event observed by one experimenter when one has in mind the map of another moving with respect to the first, there is no difficulty at all, for the second experimenter is just finding a place in his map for the event which the other observer tells him about—a situation with which we are familiar in mapping the results of geographical exploration. Consider the other extreme where a light signal sent out from a star is observed by two men in relative motion with respect to each other. On what grounds are the two distinct occurrences of the reception of the signal in different measuring instruments to be connected by the emission of the same signal? Unless the signal carries in its form some recognisable features, just as does a signal by a lamp in the Morse code, the only way we have to check that the signals came from the same source at the same local time of the source, is to show that the event which we call the emission of the light signal is represented at corresponding places in the space-time maps of the two observers; that is, if one observer gives the space and time coordinates in his map and these coordinates are transformed into those for the map of the other observer, the place so determined in the second map agrees with what the second observer has represented in his map as the origin of the signal on account of his own observations. Such comparison is impossible in dealing with atomic events. If one observer has a recording apparatus which shows the arrival of a light quantum at a particular instant, it is clear that no other observer could have detected any effect whatever at that point-instant of space-time by means of another apparatus, because one apparatus excludes the possibility of the other. The expression, 'one can mark a possible point-instant in space-time by means of objects belonging to different observers' is not an allowable one and we do not require it for the theory of relativity, if we recognise that a certain crudeness of description must be allowed. By means of different objects two observers mark what is represented as the *same* place and time in a space-time map. Consequently a plurality of observers requires a certain crudeness of the descriptions that can be made of events 'observed by them in common', and this has generally been overlooked, because the degree of crudeness seems to be dictated only by the limits of our ability to make refined apparatus.

Of another character is the uncertainty in specifying the exact transformation from one system of reference to another. If we have to transform to axes fixed to a system of small mass, then Heisenberg's principle sets a limit to the precision with which the relative translational velocity can be specified. Likewise if the moment of inertia of the system is small, the relative angular velocity will be uncertain. Whenever we have to deal with a system fixed to objects which lie near to each other in space, then on account of the limited precision of any actual method for measuring the positions of these objects, geometrical and therefore kinematical and dynamical descriptions given with this system as base are subject to uncertainties which increase in magnitude as the spatial extent of the reference system is reduced.

CHAPTER VI

SOME ASPECTS OF THE SYMBOLISM OF MECHANICS AND ELECTRICITY

IN the preceding pages it ought to have become clear to the reader what is meant by the statement made in the preface to this book that the philosophical problems connected with physics are essentially logical problems. They are to be elucidated without referring to theories of psychology and without the aid of alleged principles of epistemology. If the exposition of these matters has been successful, the experience of thinking about them should have involved the kind of feeling which one has in dealing with ordinary everyday affairs in a matter-of-fact way. In the light of such experience, when one faces a new problem, although one may not know the proper logical elucidation of it, one is familiar with the kind of process by which it is to be handled, namely, by examining our use of symbolism; one must be cautious, however, not to dismiss a problem with an appearance of generality and an amount of confidence unsupported by detailed examination. The examination need be no more impressive than the search by an electrical engineer for a fault in a transmission line or the overhauling of the engine of a motor car by a mechanic. The importance of success in the examination is not intrinsic, but is derived from the value we attach to the activities whose interruption produced the need for elucidation. On this basis we shall deal in a discursive manner with a number of topics—some well known to teachers of physics, others chosen because of the writer's interest—emphasising points in the symbolism.

What is the logical status of the dimensions of a physical quantity? Sometimes one has wished to deal with the measured length 'thirty centimetres' for example, according to the scheme 'quality, length: quantity, 30', and to say that the dimensions of the physical magnitudes which appear in any formula of physics represent the quality of the magnitudes represented in the equation. The expression '30 cm.' is made analogous to the expression '30 apples', whereas a better but still incomplete analogy would be '30 crates of apples', and the translation of 30 cm. into 12 in. would correspond to the translation of 30 crates into 3000 lb. One is tempted to make length or distance correspond to apples in the analogy and to imagine that the dimensions of the physical quantity refer to the physical nature of the magnitude; indeed, this idea is a source of actual confusion in connection with electrical units, for some men have wished to present them according to the scheme that electric charge in the electrostatic and electromagnetic systems must have the same dimensions because it is the same physical quality, electric charge, that is being represented. In order to show the error in this point of view, let us ask the question: Suppose that one adopted the lb.-ft.-sec. system for the definition of electrical units in the two systems, would the ratio of the electromagnetic to the electrostatic unit of charge have to be changed or not? Clearly instead of the number 3×10^{10}, we should have $1{\cdot}0 \times 10^{9}$.

In elementary mechanics we express the connection between a force f and the motion of a body of mass m caused by it, by writing $f = kma$, where a is the resulting acceleration and k is a number which depends on the units in terms of which the other three quantities are to be measured. Suppose that for some practical reason we wish to find the form this equation must have when we transform to a new system of units. By means of the dimensional calculus we obtain $K = F/MA$, which means that if the units of force, mass and acceleration are changed in the ratios $1/p$, $1/q$, and $1/r$, the numbers which represent the measured force, mass and acceleration respectively will be changed in the ratios p, q, r, and hence the number k will be altered in the ratio p/qr. If, however, the units of force are not chosen to be independent, but are made dependent on the units of mass and acceleration according to the law $F = MA$, then the number k will not change when the units of mass and acceleration are altered, and the quantity k is said to be dimensionless. The dimensions of a physical quantity determine the way in which the number representing its magnitude is altered on transformation of the units of measurement. It is characteristic of the method of physics that we express its laws in forms which are invariant with respect to such transformations. This choice is a matter of practical convenience in our symbolism and is not the profound philosophical principle that the laws of nature being independent of us must be invariant with respect to the units chosen for measurement. Indeed it is striking that some of the numbers which appear in our formulae as 'constants of nature' do depend on the units of measurement. For instance, Planck's constant, the charge of the electron, the constant of gravitation, all have values which depend on the units chosen, and the laws $E=h\nu$, $f=Gm_1m_2/r^2$ are left unchanged by transformation of units only if h and G are subject to change.

In electrostatics, the question arises as to the dimensions to be assigned to dielectric constant. Some men argue with force that dielectric constant is defined as the ratio of the capacities of two condensers and must therefore be dimensionless: it is then required that capacity have the dimensions of length, and this result appears to cause discomfort to others who see all too clearly that an electrical capacity is measured electrically in a different way from a length. It may be pointed out, however, that in the simple case of a parallel plate condenser the capacity is calculated as the ratio of an area to a length, both of which can be measured by means of a foot-rule. Compare this with the determination of the cross-sectional area of a narrow tube by measuring the length of a mercury thread which is afterwards weighed in a balance, or with the determination of the wave-length of light by means of a diffraction grating. There is no logical objection to assigning the dimensions length to capacity in electrostatics, for the dimensions have no reference at all to the kind of measurement that has to be used in particular instances; the dimensions refer only to the representation of measurements in mathematical equations. The equation itself exhibits the dimensional relationships which are required in calculations of the effects of change of units.

The system of dimensions of electrical units in which dielectric constant in the electrostatic system and magnetic permeability in the electromagnetic are made dimensionless was adopted by Maxwell. Others have used a system in which dimensions are assigned to the quantities just mentioned; the corresponding notation is allowable provided that we do not err in arbitrarily requiring the ratio of the electromagnetic to the electrostatic unit of charge in the c.g.s. system to be dimensionless. This ratio is no more dimensionless than that of the centimetre to the second regarded as units of distance. Let us for a moment consider the system of measuring distances in terms of the time taken by light to pass over them. Velocity then becomes a dimensionless quantity being expressed in fact as a fraction of the velocity of light. In this system the ratio of the electrical units would be dimensionless, and for this reason some men may wish to adopt such a notation; but to attempt to justify the notation on the ground that the ratio in question *must* be dimensionless is without sense.

The term force in mechanics probably has given rise to more logical difficulties than any other. In spite of Mach's illuminating historical survey of the development of mechanics and of Hertz's acute analysis of the foundations of the science in his *Principles of Mechanics*, one still finds confusion about essential things. This confusion has nothing to do with the application of mechanics but is a logical matter which arises from lack of understanding symbolism. The main difficulty lies in using the equation $f = ma$ to define the units in which force is to be measured, and may be put in the form of the question, How can Newton's second law really be a law of nature when, indeed, the equation by means of which it is expressed, appears to be a definition requiring no experimental proof whatever? Or again, How can we define mass without recourse to the second law? In order to elucidate the matter, let us recall that as a historical fact the measurement of forces in statics preceded the invention of dynamics. So long as we have in mind such an independent method of measuring force, the law $f = ma$ has to be established by experiment. Once this has been done, the ladder by which we climbed is knocked down and we rest on different ground. We postulate the law as the basis of calculation from our measurements, giving up the former methods of measuring forces in favour of the very much more precise ones that become available to us when we use dynamical laws.

The direct measurement of acceleration involves measurement of lengths and times, requiring reference to standard scales and clocks. But the operations which actually have to be carried out are not nearly so simple as this scheme of presenting them appears to suggest. Whenever a precise measurement is required in the laboratory, it may be necessary to introduce corrections which involve Newton's laws of motion—and this applies also in comparisons of masses by means of the chemical balance. We may use the law $f = ma$ to determine any one of the quantities appearing in it in terms of the other two which are measured. Consider, for example, a determination of the mass of an oil-drop in Millikan's experiment to measure the electronic charge: the force is found by Stokes's law from the terminal velocity of fall, the acceleration which must be known is the 'acceleration of gravity', and this is obtained from the period of oscillation of a pendulum and so on. Once we understand the *system* of dynamics, we are not compelled to treat the subject like a one-way artery of traffic, as so-called logical presentations of it seem to require; and the same applies to geometry. It is *impossible* to define force and mass without at least implicitly introducing Newton's laws. Suppose that we define the relative mass of two particles as the inverse ratio of the accelerations they produce in each other under their mutual action only. This definition is equivalent logically to $f_1 = m_1 a_1$, $f_2 = m_2 a_2$ (second law) with $f_1 = f_2$ (third law). It is only when we consider the definition of the experimental processes by which force and mass are to be measured that the question of precedence enters. Mass is to be measured in units derived from comparison with the standard pound or kilogram. It is this convention that renders the units of force secondary, because once the unit of mass has been fixed, there is no choice regarding the unit of force. But so far as the mathematical law is concerned, we are quite at liberty to proceed otherwise if we so wish.

In order to remove any doubt about this question it is perhaps helpful to look on the application of an entirely different law which has received little attention from philosophers and therefore as a piece of symbolism has been left fully exposed. Consider Ohm's law in electricity. The experimental fact which this law expresses is the constancy of the ratio of the voltage (E) between the ends of a metallic conductor to the current (C) through it under constant physical conditions. This ratio is called the resistance (R) of the conductor, and $E=CR$. In the absolute systems, the units of E and C are already chosen, and Ohm's law fixes the corresponding unit of R, but that does not mean that we can use the law only to find R in terms of E and C. In a certain sense we may regard Ohm's law as analogous to $f=ma$, with E, C and R corresponding to f, a and m respectively; if we think only of potentiometric devices for the measurement of currents and resistances, the analogy is a good one, for in these we may regard E as a middle term connecting two current-carrying conductors, just as we regard force in dynamics as a middle term between two motions. When we come to consider the international units, we find that for practical reasons the fundamental units are those of current and resistance; accordingly the international unit of voltage is a derived unit fixed by Ohm's law. Of course, it is convenient to define a standard of voltage in terms of standard cells, but it is a secondary standard only. It has to be remembered that the application of Ohm's law in the laboratory is by no means the simple affair which the algebraic formula may suggest. In any precise measurement it is necessary to refer to a large part of electrical knowledge—variation of resistances with temperature, pressure and age, thermal e.m.f.'s in the circuit, effect of magnetic fields on electric meters and so on. Neither is the application of $f=ma$ a simple matter in precise mechanical measurements. Once we have been shown how to apply these equations in the laboratory and see how we use them first in one form and then another—$f=ma$, $m=f/a$, $a=f/m$—we are unlikely to insist on treating the equation in one way only. Yet when it is a matter of presenting the subject there is always difficulty as to order of presentation and as to definitions.

When we adopt one scheme of presentation we should keep in mind that we could as well have adopted another. That which follows the historical development of the ideas is of unique importance however, for it compares the system now adopted with another one in which the laws appear open to experimental verification. The definition of acceleration by kinematic operations seems clear enough, nevertheless one can easily propose experimental situations where the kinematic definition cannot be directly applied with any precision and where therefore in the laboratory we should have recourse to dynamical principles in order to measure it precisely; the acceleration of gravity is in fact treated in this way. It is well to point out also in this connection that even the kinematic definition does not require us literally to measure rate of change of velocity; we can pass by mathematical operations to the relation between distance and the time, verify whether the experimental law agrees with the form of the appropriate theoretical one, then pick out the value of the acceleration as a numerical coefficient in the equation. Even if for pedagogical purposes one admits the kinematical definition of acceleration as fundamental, one cannot escape the necessity for introducing force and mass already related by the law we have in mind. The definition of mass to which we have already referred by means of relative accelerations has no logical precedence over the definition of forces as proportional to the accelerations they produce in the same body. It may seem that since the units of force are derived from those of mass, length and time, it is logically to be preferred that we should write '$f=ma$ (definition)'; but when one examines the matter more closely one finds that the operations for comparing masses precisely use this very law in the calculations. The logical point we have to keep in mind is this; when we wish to introduce ideas whose connection is represented in a mathematical law, we cannot first introduce the ideas and then impose the law on the symbols representing the magnitudes involved, for until we have the law the ideas are not made clear and definite. The numbers of arithmetic are not *entities* on which the laws of arithmetic are imposed. It is ungrammatical to treat the numbers of arithmetic as if other laws were possible, because apart from the laws the numbers have no logical properties at all.

We shall pass over the logical analysis of the measurement of mass and discuss neither the relativity of mass nor the distinction of inertial from gravitational mass. Newton's laws of motion and their place in defining a special frame of reference have been examined so often in connection with relativity that it seems unnecessary to reopen the discussion here; but all these matters are logically important in mechanics.

The description of any dynamical phenomenon is always relative to some system of reference. From the point of view of kinematics it is indifferent which system we use, for we can immediately translate the description relative to one system into that relative to another, merely by applying the geometry of space-time which supplies the rules for such transformations of coordinates. If the transformation we have adopted does not give the correct description, we have to assume some physical effect on our measuring apparatus or to change the laws of transformation, that is, adopt a different geometry. Dynamically, however, we are concerned with descriptions according to the laws of motion, and translation of the description from one system to another is possible only if we know also the laws of transformation of dynamical quantities or of the laws of motion; the latter can be deduced from the former and the kinematical laws of transformation. If we postulate the complete system of transformation equations, then observed incorrectness of the descriptions derived by transformation will require us to invent physical properties in respect to which the two frames of reference differ. The same considerations apply to the transformation of electrical magnitudes and of the laws of electricity. If we lay down the principle that the laws of dynamics are to transform covariantly under certain kinematic transformations, this does not exclude the possibility that the result of the transformation may not agree with observation. We can then say *either* that we have used the wrong kinematic transformation *or* that we must introduce physical causes to account for the difference. The choice that lies open to us is the choice of our method of representation.

We have already referred to the possibility of eliminating explicit representation of forces so far as the interaction between the members of a dynamical system is concerned by expressing the connections between the motions immediately by means of the third law. But in general it is still necessary to employ the conception 'force on the whole system'. By including enough objects in the representation we arrive at the idea of a closed system on which no force acts. This is practically realisable if we require not the theoretical vanishing of this force but only its diminution below the limit of error in measurement. For instance, no physicist would think of taking into account the gravitational effect of the table and chairs in his laboratory on the motion of the bodies he observes, nor the effect of an electromagnet in Tokio on a compass needle in Manchester. The description of his experiments does not require any such references. But he might have to take into account the radiation field of a transmitter at the Antipodes.

The connection between motions which is represented by a force according to Newton's laws is a causal connection. The word 'motion' is used here to denote, roughly, a body moving over part of its path with a certain speed—and this idea can be made more precise by taking it to mean dynamical state, i.e. a position in the space of the coordinates used to specify configuration, together with a possible set of values of the momenta. To illustrate what is meant by connecting motions, imagine an arrangement in which a small object is moved behind an obstacle to our view and that a similar object is observed at a later time to come from behind the obstacle. Ask the questions: Is it the same object or merely another like it? What motion takes place behind the obstacle so that the object emerges to view with its observed velocity? Whatever law we propose for the motion behind the obstacle would be a connection in our thought of the two motions.

In the above sense we connect causally (*a*) motions 'over the same spaces at different times', (*b*) contemporary motions 'over different spaces', or (*c*) any motions which are neither contemporaneous nor over the same space. The causal connections (*a*) and (*b*) were thought by Hume to be the only methods by which necessary connections between things are established, but they are merely the most simple methods of connection; they are used in the following way. In (*a*) the 'same space' traversed in the motions refers to objects with respect to which the motions are measured, and to all other bodies which it is relevant to mention in describing the motions adequately. Such motions are motions under the same forces, and any motion which does not follow the chosen law is motion under altered forces; it may be possible to describe this in terms of bodies which are said to affect or disturb the motion. All the motions under the same law of force resemble each other in respect of this law and are classed as similar motions.

The connection of contemporaneous motions (*b*) appears in its most obvious form when we deal with collision phenomena. One moving object traverses a portion of its path contiguous to a part of the path of another body at the same time. We learn in our earliest experience to connect two such motions because the bodies pass close to each other or collide. We say that in the collision of A with B, A causes a change in B's motion and B a change in A's. In so far as it is not necessary to refer explicitly to other objects C, D, ..., in describing the collision, we can form the closed system A and B. This requires Newton's third law to hold, otherwise the definition of force would be contradicted; the motions obey the conservation of energy and momentum. Any system to which these conservation laws do not apply is not a closed system.

We have just examined the dynamical description of those motions which by resemblance or contiguity of objects are connected in the exact necessary form prescribed by Newton's laws. We need not, however, restrict ourselves to these principles, we may evidently use any principle or law that we find useful in order to connect the motions; but whichever we apply, it is the principle itself that supplies the connection. If we care to look on the matter in this particular way, we connect, for instance, the motions of drops of differing masses and electric charges between two parallel plates by means of the electric field between the plates. When the field changes between one observation and another we must also take into account the changes in the position of the needle of the electrical instrument used to measure the voltage between the plates. Since we are free to choose the principle of connection, we have to dismiss the metaphysical controversy regarding 'contact action' and 'action at a distance' theories as a pseudo-problem.

Let the reader now stop to consider what we have been doing in looking at dynamics in this way. Instead of treating the subject as a method of representation by means of which we propose propositions to be compared with reality, we regard the experimental facts as the data and the method of representation as a system for connecting the facts which would otherwise remain unrelated to each other except by accidental external connections, as opposed to intrinsic logical connections. Compare this with the use of connected signs in a language as opposed to meaningless arrays of signs which have no application in language. Of course, it is just not true that the facts are unrelated to each other. Would the experiments have been performed without the system of ideas? Nevertheless it is of value to look on dynamics in this upside-down fashion, because when we have attained the new point of view we shall get rid of certain preconceptions as to the logic of physics. Generally speaking, philosophy requires us to examine language not from the point of view of one scheme only, but also from every other point of view we are able to occupy. This activity is a particularly trying one for mathematicians who are accustomed to settle any mathematical problem in a single proof and are not interested in a variety of proofs. Yet variety of proof is always logically important, just as variety of route is geographically important.

Whenever the velocity of motion with respect to a frame of reference is not uniform, the description in terms of Newton's laws introduces forces to represent the varying character of the motion at different places and therefore connects these parts of the motion at different places. In other words, to specify the forces presupposes that the motions at different places are integrated into a single continued motion. A particular specification of the forces represents that the integrated pattern of the motions at different places follows the proper law under these forces. But the observed facts need not have the form of a number of dynamical states specified at arbitrary times. There may be no local observation of velocity at all; all the observations may be measurements of angles, distances and times as in the observed motion of the planets in the heavens, and velocity may be deduced from these particular observations by means of the laws of motion.

Our schematisation, put forward purely for the purpose of simple illustration, treats dynamics according to the pattern of the mathematical problem of drawing a curve satisfying a certain differential equation to pass through a given set of points. But the data sufficient to determine the curve might have been given in an entirely different form in which there is variety in their geometrical nature. One can, for instance, determine a circle to pass through three given non-collinear points, or through one point and touching a given straight line, or touching a pair of lines and so on. In physics our experimental data have this kind of variety, and any scheme which ignores this fact is an inadequate representation of what is done by means of physical symbolism. With this reservation regarding the scheme we are using, we can still employ it to show some aspects of the function of physical symbolism that are not easily open to inspection without it.

Motion is relative to particular objects, so that when different specifications of force are required in describing two similar dynamical phenomena, we ascribe the different forces to the different objects (or states of objects, e.g. electrification) which are relevant to the description. On account of the experimental possibility of making small closed systems, according to which when a body is removed to a sufficiently great distance it need no longer be taken into account—we cannot abolish things in the way we leave them out in our theories—we can investigate the effects of individual objects and ultimately specify the force due to each of them not merely along one orbit, but at all points of space relative to the object which causes the force. The connection of motions by means of force at a point of space is an infinitesimal connection along an orbit; finite connections are set up by means of the potential energy. It is clear that by potential energy other dynamical properties of bodies in addition to their mass are introduced into dynamics. The change in the kinetic energy of a body in the course of its motion is expressible as a function of space only, provided that no work is dissipated in heat, sound, light or other electromagnetic effects: this, Hertz pointed out, amounts to requiring that all motions of the closed system shall be specified in the equations:

> The difference between conservative and dissipative systems of force does not lie in nature but results simply from the voluntary restriction of our knowledge of natural systems. If all the masses of the system were considered visible, then the difference would cease to exist and all forces of nature could be regarded as conservative forces. (*Principles of Mechanics*, § 665.)

It may be pointed out, however, that in achieving such a form of description we may need to invent new entities. Mere assertion of lack of conservation is useless for physics: we have to give the law of departure from conservation under the different circumstances to which this law is to be applied, and this is equivalent to conservation with all motions taken into account.

So long as we concentrate our attention only on single orbits, we have a limited view of the connection of motions by means of the potential energy, for this function serves also to connect with each other any two members of the family of orbits described by the same form of the potential energy (and of course of the kinetic energy as a function of the momenta or of the velocities). The way in which we propose such a family of orbits, or system of possible motions, for study is to specify either the Lagrangian or the Hamiltonian function.

Whenever we have to deal with bodies which can be treated as particles, that part $V_{A\alpha}$ of the potential energy of the motion of the particle α which is due to the presence of the particle A, depends on the relative coordinates of A and α. We split $V_{A\alpha}$ in such a way that one part refers to A and the other part to α, by inventing the idea of specific potentials such as gravitational and electrostatic potential. $V_{A\alpha}$ is converted into $\phi_A(\alpha)$, where α is now regarded as a *variable* whose dynamical particularity is not represented in the form of $\phi_A(\alpha)$, although the spatial coordinates of α are mentioned. We may therefore suppress α in the mathematical formula, and this is actually done in mathematical physics. Nevertheless, whenever gravitational or electrostatic potential is used in calculations in mechanics, α must be restored and given a particular value.

Any particular system of motions conformable with a given Hamiltonian function is internally connected in a way that has been given explicit mathematical expression by means of the transformation theory of dynamics. When motion is represented by the canonical equations, the transformation from the values of the dynamical variables at time t to their value at the later instant denoted by an infinitesimal increment in t is an infinitesimal contact transformation. The series of contact transformations by which the series of values of the variables at successive instants of time is generated, may be pictured as the unfolding of a single transformation process, and the trajectories of the system in the space of the configurational coordinates bear a relation to this process analogous to that between the rays of an optical system and the propagation of the wave front. This analogy has been described in detail so frequently in works on quantum mechanics that there is no need to repeat its description. In order to emphasise this matter of connection of motions, it is more profitable to point to the place in the theory occupied by any integral of the canonical equations as a generating function of those contact transformations which transform any orbit into an adjacent orbit. Thus instead of connecting motions only along single orbits as in the simple Newtonian scheme, the Hamiltonian connects motions by a principle which involves all* the orbits; in fact any possible motion of the system *can* be connected with any other part by means of a contact transformation. In the classical method, only the transformations along the orbits correspond to physical processes. So far as other connections of the system are concerned infinitesimal transformations are always possible. The quantum theory in taking over the Hamiltonian method has imposed restrictions so that infinitesimal transformations are no longer always possible, because the particular laws of discontinuity governed by Planck's constant have been introduced. Of all the generalised systems of dynamics that we know based on Newton's laws, that in the Hamiltonian form is the only one which explicitly connects the orbits by transformations, and thus makes it possible for us to imagine that finite transformations of this type may represent possible physical processes. All the other formulations of dynamics rest on connection along the orbits by continuous motion of the system.

Without intending to deal in any detail with the methods of quantum mechanics, we may look at a few of its ideas which have outstanding logical interest.

Consider the Hamiltonian which defines the system of a particle A, such, for instance, as an electron in a field specified in the form of H. By solving the equations of motion, we describe the motion not merely of one particular object A_1 in this field but also the motion of similar bodies A_2, A_3, ..., provided that the motions are not contemporaneous. That is, although H in the form given can represent the motion of any one of the A's separately, it does not represent the contemporaneous motion of two or more of them. In order to represent this, it is necessary to use a Hamiltonian in which each

* There is of course no totality of orbits any more than there is a totality of cardinal numbers in arithmetic.

A is represented by a complete set of dynamical variables. If there are n objects, H is the sum of n exactly similar functions; such a system is said to be degenerate, and the possibilities of motion are the same as for only one particle in the given field. There is nothing in the solutions except the suffix r attached to the coordinates of A to distinguish the objects $A_1, A_2, \ldots, A_n$. So long as we have a rule by which to assign the suffixes, we may say that we represent the motion of n objects; but in the absence of such a rule the solutions of the equations of motion must themselves show the multiplicity of objects, and this means that the degeneracy of the system must be removed by introducing into the Hamiltonian the potential energy of interaction of A_r and A_s $(r, s = 1, 2, \ldots, n)$. One is led to ask, if we can represent the motion of any one A in the field by a given H, can we not represent any number of A's contemporaneously moving in the field by a Hamiltonian in which the number of similar objects is regarded as a variable which can take only integral values? In this way we should group together by the rules of our symbolism, systems of $1, 2, \ldots, n$ particles. This type of representation is actually made in modern physics, but the variable number of objects is interpreted statistically.

In dealing with a system of similar particles which are distinguished by their dynamical coordinates only, if we think of the motion as the transformation from the initial set of coordinates to the final set, we face a difficulty. According to the classical scheme the coordinates of the particles in the initial and final states are connected uniquely by the motion of the individual particles. There is one-to-one correspondence between the sets of particle coordinates in the initial and final states, and this is denoted by the use of suffixes. The law of variation with time of the coordinates (x_r, y_r, z_r) gives the motion of the rth particle. Now, unless outside the symbolism considered here some rule is given by which the suffixes can be assigned to the particles, there is nothing but the hypothesis of continuity to decide how 'r' shall be transferred from one particle position to another, and not give place to the suffix s; and even this hypothesis is not adequate when collisions occur, for it is then necessary to have another rule by which r and s shall be assigned after a collision. It might appear that the rule for assigning the suffixes throughout the motion of the system would be introduced into the mechanical symbolism by taking into account the interaction of the particles with the detecting system, but this is not the case, for exactly the same problem would arise with regard to numbering the effects of the interaction; we should have to depend on the hypothesis of continuity. The particles must be distinguished by means of a label which is not variable like the coordinates. The classical picture of similar particles involves therefore that there should be some rule for assigning suffixes which is equivalent to labelling the particles. Now physically such a rule would require the effects by which the labels are to be recognised. This is all right if we are dealing with the motion of a system of coloured balls which we can see, but with a system of atoms or electrons we are not permitted to have labels of that kind, hence the motion is to be looked on as the transformation from the initial *set* of coordinate values to the final *set*. In the calculation the suffixes are used to distinguish variables; when the calculation is completed, the particular values they take are distinguished by their being different. Numerical suffixes are not necessary, we could use any kind of sign we like to choose for the purpose of distinguishing; the kind of sign is not determined in the calculation.

If we are dealing with a system in which, for physical reasons, labels are not permitted, unless the hypothesis of continuity can be applied, the classical method of description fails. This failure is overcome in quantum mechanics by considering interchanges of labels as motions allowed in the system. Such an interchange is pictured as a physical process, but evidently its possibility arises only in the symbolism which makes use of labels and then has to remove the restrictions introduced by their use. Mathematically we may express the matter thus: A transformation of a set of n points in space into another set of n points may set up a one-to-one correspondence between the points of the two sets, but if this correspondence is not demanded, the transformation is not unique; there is a set of possible transformations which are all equivalent, and which are related by means of the group of the transformations any one of which transforms one or other of the sets of points into itself. A geometrical illustration is the transformation of a closed curve into another closed curve near it, the two being connected by an element of tube. For one-to-one correspondence of points by the transformation, the connection would be represented by a particular family of line elements on the element of tube. For the infinitesimal transformation of one curve into the other without this correspondence, the connection is represented by the tube surface. The idea of the motion of a system of points in the above sense may therefore be useful for generalising mechanics.

In trying to understand a theory we are sometimes tempted to immerse its symbolism in a more general scheme or theory (G), believing that by so doing we explain or clarify the theory (a) with which we started. The scheme G is relevant only if it points to other possibilities b, c, ... for representing nature; that is, if we say that a is the correct theory since b, c, ... do not apply to the facts, then of the group a, b, c, ... the only theory that is relevant to physics is a. Any law given to select a from G as the correct theory has no place in physics because the system G of representation is not a useful one until its possibilities are found to correspond to possibilities in nature. To be practically useful, a symbolically correct generalisation must add to present knowledge new possibilities and lead us to look for them. If they are not found then the generalisation has to be dismissed.

Of a different type are those generalisations of physical laws which work by our passing from systems of possibilities expressed by simple laws to others given by more complicated laws. For instance, in dynamics let us consider the Hamiltonian function for the motion of a charged particle of given mass (m) and charge (q) in a given electromagnetic field. For each pair of numbers m, q there is a distinct dynamical system. Let us make m and q variable and regard as belonging to a single new system of possible motions the possibilities defined by any of the corresponding dynamical systems. In this way we have arrived at a more general idea of a system of possible motions, but the dynamical systems on which it rests, being picked out according to the law of the given electromagnetic field, are very special systems. We choose a Hamiltonian function of a definite form; by making variable some elements in that form (m and q above) we get a system of Hamiltonian functions, and a mathematical expression of the intrinsic form of this system may be obtained by using some process which eliminates the variables from the expression for the Hamiltonian functions belonging to the system. In illustration of this, consider how the analytical equation for a circle in a plane involves the radius of the circle and the coordinates of the centre, and how by differentiating the equation we can obtain enough relations to eliminate the variable parameters, so obtaining the differential equation of families of circles. Roughly speaking, we may say that the differential equation expresses the circularity of each curve of the family. In the mechanical question we have been examining, we may say that the property which distinguishes the *system* of Hamiltonian functions is the given electromagnetic field. We may therefore speak of this system of possible motions as an electromagnetic system. In this way we have shown how electrical symbolism *connects* a system of particle motions of more complex possibilities than ordinary dynamical symbolism. Since the connection is one of motions of a particle, we may quite properly regard electrical symbolism as part of mechanical symbolism. We shall have occasion again to refer to this way of looking on electricity.

The idea of electricity arose from comparisons of the motion of bodies in the absence of electricity (and magnetism) with the motion of the same or similar bodies in the presence of electricity. All electrical quantities are defined through the laws of electricity in terms of the mechanical effects by which we can measure them. In order to define a quantity such as electric charge or magnetic field, we require to connect at least three distinct motions. For instance, one (A) may be motion in the absence of electrical effects. To obtain the mechanical effects of the given charge and of the standard unit of charge, we compare each of the other two motions with *A*. The value of the charge or field is deduced by comparing these mechanical effects. This threefold multiplicity of connection between motions distinguishes the electrical connection of motions from that given by the potential energy in dynamics where, according to the same method of analysis, there exists only a twofold multiplicity.

The facts of electrodynamics say that some of the motions and arrangements of particular pieces of matter produce electrical effects. Consider such a simple phenomenon as the connection of the poles of an accumulator to the terminals of a voltmeter by means of copper wires. Without an electrical model, according to ordinary mechanics, this connection of things should produce no mechanical effects of the kind observed; but the voltmeter coil is turned.

In describing an electrical phenomenon we have to specify the configuration, chemical constitution and physical state of the objects participating in it. The chemical specification is achieved through physical properties determined by mass, length and time measurements and also by measurements of derived quantities such as pressure and temperature. Whenever questions of purity arise, electrical and optical measurements are required also. Of course the science of electricity made its early progress with the aid of crude notions of materials and of the ideas 'conductor', 'insulator', 'magnet' and so on; but with the advance of knowledge much more refined descriptions have been introduced, so that the whole structure of our knowledge involving electrical symbolism is extremely complicated, and its logical character is certainly not to be laid bare by simple schematisations intended to have general application.

The electrical theory which is summarised in Maxwell's equations and the conventions of electrical measurements should, I think, be distinguished from the electrical theory of matter by means of which we show how to calculate on certain hypotheses the measured electric and magnetic, thermal, optical and mechanical properties of chemical substances, and of any naturally occurring or manufactured arrangements of them. Whereas in the former theory, dielectric constant, etc., are taken from laboratory measurements and used in the equations, in the electrical theory of matter we are concerned with explaining why a particular substance or piece of it should have its observed properties. When we wish to measure the properties of a piece of matter it is made part of an electrical (or magnetic) circuit or placed in a field. Either we use light to examine what happens inside or we deduce this with the aid of Maxwell's equations from the boundary conditions given by electrical and mechanical measurements. In electrical measurements, Maxwell's equations play a rôle analogous to the rôle of Newton's laws in the measurement of mass, for example. It is customary for books on electrical theory to present the theory of circuits before Maxwell's equations, and this follows in some, but not all, respects the historical development of the subject. Now that we have reached the position that Maxwell's equations are the final appeal in all theoretical questions connected with electrical measurements, they should be given the logical prior place, and the laws of circuits should be deduced from them as integral properties of the field under the conditions which make the experimental circuit laws valid.

In practice the electrical circuit is of fundamental importance, for it represents how its parts are connected and in an accurate specification gives their geometrical configuration and, if necessary—as in electrical machinery—their motion also. An object may appear in the circuit merely as a resistance, capacity or inductance, and so on. But, however we describe it, all the quantities which we use in the description can be defined in terms of Maxwell's equations (or the circuit laws), operations with electrical measuring instruments, scales and clocks, and all other relevant laboratory procedures such as are specified in the definitions of the international ohm and ampere. The electrical circuit shows how far the system extends—any particular object is either to be included in it or it is not, but it is not always possible to make this kind of description; such cases are covered by the conception of radiating systems. We are never concerned with an electromagnetic field where we cannot measure its effects; if we do not put objects in the field to measure it by its mechanical or optical action we calculate the field strength by Maxwell's equations from the boundary conditions, as, for instance, in e/m deflection experiments.

Given the properties of the materials which they are to use —these can be measured—the electrical engineer and experimental physicist, in actual practice with the aid of electrical theory, design their machines and apparatus to achieve whatever ends they have in view, just as ordinary mechanics is applied in civil engineering. These systems work because the laws of the theory being postulated in getting the data cannot be called in question by means of these data. Any facts which do not fit the laws will be dealt with by additional hypotheses. This is a matter of the conventions of physicists analogous to those in the use of Newton's laws.

When we are concerned with the calculation of the electrical properties of matter by the methods of atomic physics, we may look on the electrical theory of matter in another way. Instead of building up the measured properties from the hypotheses of the theory—in many cases with superfluous refinement in calculation—we may use these properties and the laws of their dependence on measured physical conditions to determine the properties of the atoms, etc. For instance, to take a well-known example, either we attempt to calculate from our theory the *F* curves obtained in X-ray diffraction experiments, or from the observed curves we calculate just that amount of atomic or molecular structure which the experimental evidence will support. The electrical theory of matter is therefore the set of principles by which we connect the measured electrical and other properties of materials.

On account of the high precision of spectroscopic measurements and the success which has attended the application of the quantum theory in describing spectra, a convention is being established, certainly many have adopted it, that atoms and molecules are to be defined primarily in terms of these optical measurements, which, apart from phenomena like the Zeeman effect, yield numbers characteristic of the atoms and molecules themselves. Optical intensities depend on the method used to excite the light, and since they are to be explained in terms of the interaction of the atoms with their surroundings and each other, are not a property of the atom by itself.

We have introduced a scheme to show how we may regard electrical symbolism as a highly developed branch of mechanics. At the same time by referring to the use of electrical theory, we have emphasised that an attempt to express in the forms of our scheme any of the representations that we deal with in the electrical laboratory will probably not succeed on account of their complexity. Imagine trying to deal with an electric motor or an alternating current bridge in this way. But the scheme is not necessary in practice, for electrical symbolism is the shorthand appropriate to handle these problems. What is important for us to recognise is that it is a shorthand, and that it enables us to manage both theoretically and for practical purposes more and more complicated physical processes. This way of looking on the theory of electricity should be opposed to the widely used view that since electricity is the stuff of which matter consists, mechanical ideas and laws should be given an electrical explanation. Let us for a moment examine this view. We have already noticed how the laws of electricity are connected with certain possible types of motion; it is possible to *conceive* of mechanical systems whose motion cannot be described at all as due to electrical forces. Consequently an electrical explanation of mechanics presupposes that, matter being electrical in nature, none of these non-electrical motions are possible. The effect of the explanation would be therefore to shut out these possibilities from consideration, and since in physics we discover only what we are able to recognise, we are likely to hinder rather than to advance the progress of physics by closing our minds in such a way.

Imagine that a mathematical physicist in the time of

Hamilton had propounded the question, How should we describe the most general possible continuous motion of a particle? There are two ways in which he might have solved his problem. The first comes from studying the Lagrangian function of a particle with respect to rotating axes of reference under a given potential energy function. Then by allowing the components of angular velocity and the potential energy any functional dependence whatever on space and time, he would arrive at equations of motion formally equivalent to those of a charged particle in an arbitrary but given electromagnetic field. The second method which could be followed is to invent the idea of potential momentum to take account of non-conservation of momentum in the same way that potential energy secures the conservation of energy. Any particle motion considered in this way will be the same as that of unit charge in a given field derived from the potential momentum and potential energy as vector and scalar potential respectively of the field. It is then a simple matter to pass to the idea of particles of different charges moving in the same field. Compare this step with the corresponding one for gravitational forces. We need not follow here the whole mathematical argument, which has been given elsewhere. The important logical point is that the idea of electricity *could* have been reached by dynamical considerations such as these. By similar studies we seem to be able to invent new symbolism for physics; it then remains to find actual processes in nature to which we can apply the symbolism and so establish its usefulness, but we must be guided by the theory where to look, just as so often in the past, and as has been so clearly shown in the recent discovery of the neutron, where the correct disposition of matter with respect to an expansion chamber made all the difference between no evidence of neutrons and seeing the tracks of recoiling atoms.

When the symbolism of electricity is regarded as advanced mechanics the threefold connection of motions which we have already mentioned, as opposed to the twofold connection in ordinary dynamics, represents in the simplest way possible the *order of complexity* characteristic of electrical symbolism.

Defined in this way the order of complexity can be associated with systems of mathematical forms to exhibit in a clear way how Hamiltonian dynamics uses up the possibilities of representation provided by one branch of mathematics and how electricity uses the next more complicated system. We shall not enter here on the technical exposition of this point of view, it suffices to make the logical point that higher orders of complexity in our symbolism are possible. Add to this the historical fact that dynamics is about three hundred years old and we are unlikely to be impressed with opinions which were expressed towards the end of last century and are now being repeated that the main theoretical principles of physics are known. Rather we shall look to the future for even greater development than has yet taken place.

When the electromagnetic theory displaced the elastic solid theory of light, optical theory was given a new life and great technical advances followed, but most of us still think of an electromagnetic field as a physical existent in the same sense as the elastic solid. To specify a field requires us to name at each point of space corresponding to each instant of time, the values of two three-vectors which represent the field. This specification is a law of transformation from the space-time of ordinary physical existence to the space in which possibilities are selected by giving particular values to the components of the electric and magnetic field vectors. Now although the word 'transformation' is a substantive, are we justified in treating this particular transformation as a thing in physics? One can indeed point to other vector fields in physics which are clearly not thought of as things. Consider the vector field which describes a particular deformation of an elastic body. Here the transformation can be represented physically by means of the displacements of the particles denoting elements of the body. But no one wishes to think of the system of displacements as a physical object. When we consider the vector field of the momentum of a particle derived from one solution of the Hamilton-Jacobi equation of its motion, we are not tempted to ask what thing this field is intended to represent. It does not represent a thing at all, only a system of possibilities for the motion of a particle in the dynamical system considered. Why then do we treat the electromagnetic field in a different way, as if it were substantial? In part, at least, the answer to this question is concerned with the dynamical theory of the field (in Maxwell's sense), the doctrine of energy and momentum distributed in space, and the idea of a system of stresses determined by the field components at each point. Guided by the analogy of the elastic solid, we think of the energy and the momentum as the energy and momentum of deformation and motion of a medium that fills space, and the state of which is specified by the field components; whereas the electromagnetic field refers to entirely different possibilities, namely, those of mechanical force on a charged particle or small piece of polarisable matter moving in it, of the electromagnetic and thermal effects in an exploring coil of wire, of the electro- and magneto-optical effects on a piece of transparent matter or on a light source placed in it, in fact, to sum up, it refers to possibilities of any of the observable physical effects which are due to the action of the field.

Even if we succeed in freeing ourselves from the restriction on thought involved in the idea of a medium to support the field, we are still faced with the problem of the apparent substantiality of the energy, momentum and stress system. This problem is disposed of by referring to the possibilities of mechanical action in the field and to the application of Newton's third law. In special relativity, which must be taken account of in considering electromagnetic phenomena, the third law retains its meaning only so far as contact action is concerned; in all other cases some hypothesis as to the law of propagation of force from one part of space to another must be introduced in order to remove the ambiguity which exists as to how the law is to be applied. In classical theory we require conservation of energy and momentum reckoned instantaneously and continuously, and we know as an experimental fact that when one body in the field loses mechanical energy and momentum, another body does not immediately gain it. The field thus appears to be a convenient invisible to balance energy and momentum accounts. But we never bring the energy and other mechanical properties of the field to light unless we place in it some object whose motion we can observe, or at least employ some apparatus or other that is mechanically influenced by the field and that can be observed. The great power of the conception of the electromagnetic field is that it allows us to leave this particular effect unspecified—to treat it as a variable. In any actual arrangement to measure the field by its effects, however, we have to restrict ourselves in calculation to the particular device we have chosen for the measurement. In brief, the electromagnetic field is a piece of symbolic machinery: one is merely overlooking the manner of its use in calculations where actual things are represented, if one regards a map in which the field is specified in the same way that one regards a map of Canada. The map of the field represents possibilities of mechanical force and optical effects in a way not unlike that in which the plan of a steamship represents possibilities for the accommodation of passengers. Actual use of the plan involves that places to be occupied are marked in it when they have been booked; it is part of the apparatus used by the shipping agent in calculations as to the stocking of the ship for the voyage and so on. The symbolism of electricity enters into physics as part only of the complete calculations which bring us sooner or later to use our hands, our eyes and our ears.

The picture of an electromagnetic field as the state of a medium is a model which dresses up the mathematical equations by which one specifies the distribution of electric and magnetic force. The advantage of the model is that in a certain sense it can be taken in at a glance and held up easily for examination. It is the substratum for the thoughts in which it is used, just as the mathematical signs we write on paper are the substratum for the mathematical connections which we express by means of them. In the preface to *Quantum Mechanics* Dirac wrote:

...nature works on a different plan. Her fundamental laws do not govern the world as it appears in our mental picture in a very direct way, but instead they control a substratum of which we cannot form a mental picture without introducing irrelevancies. The formulation of these laws requires the mathematics of transformations.

And here the word 'substratum' appears to have a meaning different from that we have used, as if there were a substratum of hidden machinery (recall the electromagnetic field) working according to laws different from those used in ordinary mechanics. 'The fundamental laws' certainly control the calculations which the mathematician makes by writing signs on paper, but what other substratum can the mathematician point to than these signs, if there is no physical model to exemplify the laws? Think of the equation $x = a \sin nt$ expressing the law of simple harmonic dependence of a variable on the time, what is the substratum controlled by this law if we do not have in mind its application to describe some physical phenomenon? In mathematical physics the signs are the apparatus for expressing laws of connection, and they are merely a part of physical symbolism. Instead of thinking of a substratum, we ought to think of the 'superstratum' of our experience of actual things and processes, into the description of which the laws enter as part of the necessary symbolic apparatus, and in this respect the new laws do not differ from the old. All that has happened is that we are now forced to call in question our former uncritical attitude to *all* physical symbolism, in order to surmount logical difficulties in understanding the new theories. But the method of physical science is the same as ever, and for proof of this statement one has only to look inside a laboratory.

In conclusion let us turn our attention briefly to the logical status in the theory of electricity of the experimental fact that electric charge exists only in integral multiples of the quantum $e = 4{\cdot}8 \times 10^{-10}$ e.s.u. We have already considered how continuity or discontinuity applies only to our method of representation. If we regard electrical phenomena as continuous the existence of the charge quantum has to be accounted for by a theory transcending electricity, because the possibility of continuous variation of charge has been postulated in the theory. On the other hand, if we regard as impossible that charges can exist different from ne (where n is an integer), then we face the type of logical problem which was encountered in regard to the introduction of the quantum of action into dynamics, for the hypothesis of atoms of electric charge is logically as much out of place in the continuous electromagnetic theory of Maxwell as is the quantum of action in the continuous dynamics of Hamilton, unless we think of the electron's charge as theoretically divisible, and that for some as yet unexplained reason, the bundles of electricity actually found in nature all have the same size. Which point of view will prevail is for the future to decide. If it happens that discontinuity is accepted, and with it limits on the precision with which an electromagnetic field can be specified, perhaps we shall be fortunate enough to avoid the mistake of attempting to use the new physical theory as a peg on which to hang 'philosophy'.

INDEX

Printed in the United States
By Bookmasters